THEORY OF LINEAR INDUCTION MOTORS

THEORY OF LINEAR INDUCTION MOTORS

By

Sakae Yamamura

A HALSTED PRESS BOOK

JOHN WILEY & SONS

New York – London – Sydney – Toronto

© University of Tokyo Press, 1972

Published in the United States, Canada, United Kingdom and Europe by Halsted Press, a Division of John Wiley & Sons, Inc., New York.

Printed in Japan.

Library of Congress Cataloging in Publication Data

Yamamura, Sakae, 1918–
   Theory of linear induction motors.
   "A Halsted Press book."
   Bibliography: p.
   1. Electric motors, Induction.   I.   Title.
TK2785.Y35   621.46′2   72-5208
ISBN 0-470-97090-1

CONTENTS

PREFACE

This book is a report on the results of research activities undertaken at our laboratory in the Electrical Engineering Department, Faculty of Engineering, University of Tokyo and is not intended as a comprehensive reference on linear induction motors. Although theoretical work started in 1964, both theoretical studies and experimental activities are still being carried out at our laboratory. Nevertheless, we felt that it would be appropriate to publish the book at this time. All the material included in the book have already been published in numerous technical articles written by me and my colleagues.

In recent years, attempts to develop new means of high-speed, efficient transportation have led to considerable world-wide interest in high-speed trains. This in turn has generated interests in the linear induction motor which is considered to be one of the most suitable propulsion systems for super-high-speed trains. Research and experiments on linear induction motors are being actively pursued in a number of countries, among them Japan. Unfortunately, many researchers, in their desire to achieve immediate practical results, have concentrated on experiments with large-scale testing equipment and large-size test trains, leaving the theoretical aspects of the linear induction motor neglected so that few useful results have been produced. In spite of extensive experimental efforts, there has been no reported test result on a linear induction motor with proven feasibility for high-speed trains higher than, say, 200 km/h. This situation is partly due to the fact that up to now no sound theoretical basis for linear induction motor has been established so that many researchers have based their ideas on theories and experiences derived from the rotary induction motor.

The essential difference between the linear induction motor and the rotary induction motor is the open linear air gap, which has both an entry end and an exit end. The end effect, which is caused by the open-endedness of the air gap, produces considerable distortion in magnetic field distribution and peculiar phenomena which are not observed in the rotary induction motor, but which considerably influence the characteristics of the linear induction motor. The first reports on the end effect of the linear induction motor were made many years ago in connection with the arch motor, an induction motor in which one part of the stator core was removed. Since then a great number of researchers studied the linear induction motor, but none succeeded in deriving solutions for the field equation of the air gap, which take into consideration the end effect and which are applicable to practical problems of the linear induction motor. As a result, the problem of the end effect was neglected and its influences were underestimated in almost all recent research and experiments on the linear induction motor, that were mostly based on the equivalent circuit theory of the rotary induction motor, which is not applicable to the linear induction motor. The misleading conclusions derived from such studies are widely accepted and have even been applied to large scale tests and experiments which, of course, might not lead to any satisfactory results.

At our laboratory, we have been studying the linear induction motor for some time and we think that we have succeeded in establishing workable theories which enable us to calculate the characteristics of the linear induction motor under the influence of the end effect. The calculations enabled us to discover several interesting, new phenomena in connection with linear induction motor theory and its practical application. If these hitherto unknown phenomena are not taken into consideration in studies on the linear induction motor, a clear understanding of its performance will be impossible and attempts to develop practical high-speed linear induction motors may be unsuccessful. The influence of the end effect is extensive, especially in high-speed linear induction motors, and it considerably reduces thrust of high-speed linear induction motors in the low-slip region. The power factor and efficiency are also reduced to a great extent. It seems rather strange that the end effect has been practically ignored in spite of its extensive influence on performance of the high-speed linear induction motors. It appears that the lack of a workable theory for the linear induction motor is partly responsible for the ignorance. The theory developed by our laboratory, and about which we have written numerous reports over a period of years, revealed the seriousness of the end effect.

After the seriousness of the end effect was made clear, it was necessary to find a way to eliminate it in order to develop a high-speed linear induction motor of practical value. We succeeded in extending the theory of the end effect to the compensation theory of the end effect and this led to the derivation of two new kinds of linear induction motors; the compensated linear induction motor and the wound-secondary-type linear induction motor. In both types the end effect has been completely removed or considerably suppressed and outstanding performance of the high-speed linear induction motor has been achieved.

Our research activities are concerned mainly with the theoretical aspects of the linear induction motor, however, we tried to demonstrate experimentally the important conclusions derived from our theories as much as our limited manpower and research funds permitted. As mentioned above, the serious influence of the end effect had been overlooked and the optimistic views prevailing did not take this influence into consideration. Under these circumstances, it was not easy to obtain help for our research activities. It was very fortunate for us that the Toray Science Foundation provided us with research funds.

Our research activities are still progressing. However, we think that there is an urgent need for the relations between newly-discovered phenomena and newly established theories of the linear induction motor to be clarified, and that understanding of the linear induction motor be promoted among interested researchers. Toward this end we have combined into one book the articles which reported our findings part by part as they were found. We believe that this step will establish a firm foundation on which further research and progress can be based. We know through technical publications that similar efforts are being carried out in other countries. We hope that the material presented in this book will prove to be useful to researchers in these countries. With this in mind we decided to publish this book in English, in spite of difficulties it presented especially since all of our original articles were published in Japanese.

My colleagues in the research project are Haruo ITO, assistant, Yoshitoshi ISHIKAWA, graduate student, and Dr. Farouk Ismal AHMED, whose contributions have been made through calculations and experimental efforts. In addition, they have provided original ideas, without which the research project would not have born fruit. I express my heartfelt thanks to them.

Tokyo
September 1972

Sakae YAMAMURA

THEORY OF LINEAR INDUCTION MOTORS

Chapter 1

Introduction

The difference between linear induction motors and rotating induction motors is basically due to the difference in their air gaps. The linear induction motor has an open air gap with an entry end and an exit end, while the rotating induction motor has a closed air gap. It is the open endedness of the air gap that gives rise to the peculiar characteristics of the linear induction motor. Other constructional features, such as larger air-gap length and a secondary conductive sheet, are not peculiar to the linear induction motor. There are rotating induction motors which have a secondary conductive sheet and there are also rotating induction motors whose air gap is larger. They are, for example, cup motors, canned induction motors and screened-rotor induction motors. Problems arising from a larger air gap and a secondary conductive sheet are common both to rotating and linear induction motors, and have been rather extensively investigated. Unfortunately, much of the research activity on linear induction motors has been concentrated on problems arising from a larger air gap or the secondary conductive sheet and not enough activity has been directed toward solving problems arising from the open air gap. Although some articles and books (e.g., ref. 1) have been written on the problems of the end effect, the effect is not clearly understood and its influences are understimated. One reason for this may be due to the fact that the end effect problem is rather complicated and very difficult to analyze theoretically. A number of authors have tried to derive rigorous analytical solutions for the air-gap field of the linear induction motor. Some have succeeded only partially, and others have failed completely. None has succeeded so far in achieving an analytical solution which would make it possible to calculate correctly the magnetic field distribution and the characteristics of the linear induction motor.

Another reason why the end effect has remained unanalyzed is that the linear induction motor has been confined to low-speed applications, the influence of the end effect in low-speed applications is not significant, and the characteristics of the low-speed linear induction motor remain comparatively unchanged even under the influence of the end effect. Under certain conditions they may be improved to some extent. We discovered that the linear induction motor is capable of producing thrust even at synchronous or higher speeds, if its synchronous speed is not high.[4,13] The characteristics of the high-speed linear induction, in contrast, experience acute and adverse influence due to the end effect; thrust is reduced, the power factor is reduced markedly, and efficiency is reduced in the small slip region, where the motor runs most of the time. The degradation of the motor performance is so pronounced that the feasibility of the linear induction motor for high-speed applications becomes questionable.

Our laboratory succeeded in deriving rigorous analytical solutions for the field equations of the air gap of the linear induction motor under the influence of the end effect. These solutions make it possible to compute the magnetic field distribution in the air gap and the characteristics of the linear induction motor. We derived three solutions, each with its own merits; the first one[4,13] is based on a one-dimensional model of the motor, the second[14] is based on a two-dimensional model, and the third[7] is based on numerical calculations. The one-dimensional solution is rather easy to derive, and yet it gives much insight into phenomenal structure of the end effect. The rigorous two-dimensional solution takes into account the two-dimensional distribution of the air gap field. This led to the development of the compensation theory of the end effect from which we conceived the compensated linear induction motor.[14,16] The compensation eliminates or suppresses the end effect and remarkably improves the performance of the high-speed linear induction motor. The third solution is a numerical calculation by means of the relaxation method. It is based on very sound assumptions and provides a true picture of the boundary conditions to the other two solutions of the magnetic field in the air gap.

The research results achieved by our laboratory on the linear induction motor will be explained in the chapters that follow. All the material in this book has already appeared in numerous technical papers and articles published by us.[2-16,19-26] They have been correlated and are reproduced here with supplementary explanations for easier understanding. The material covers research activities from 1964 to 1972; research on the subject is still in progress. Inclusion of results from research other than ours is limited to those which are absolutely necessary to the understanding of our theories.

SUMMARY OF RESULTS

(1) *Solution for the Field Equations of the Air Gap*

During our studies we found that no solution of any practical value existed for electromagnetic field equations of the air gap in the linear induction motor. The lack of such a solution was the reason for so much confusion and lack of progress among researchers throughout the world who were engaged in the development of high-speed linear induction motors. In our laboratory, we derived three kinds of solutions for the field equations. (a) The one-dimensional solution which takes into account one-dimensional field variation of the air gap in the direction of the motion of the motor. (b) The two-dimensional solution which takes into account the field variation of the air gap both in the direction of the motor motion and the direction perpendicular to it. (c) The numerical solution using the relaxation method. The three solutions each have their own merits in certain ranges of applications to practical problems.

(2) *The Nature of the Travelling Magnetic Field in the Air Gap*

Although the existence of the end effect was an accepted fact, details of its structure were not known. We were able to determine these details using our one-dimensional and two-dimensional solutions. Our studies showed that the entry-end-effect wave is produced by the entry-end effect and travels with slower decay in the same direction as the normal travelling wave at synchronous speed, while the exit-end-effect wave is produced by the exit-end effect and travels with quicker decay in the opposite direction. We also derived analytical expressions for the end-effect waves which give amplitudes, speed and lengths of penetration of the end-effect waves as functions of various parameters, such as motor speed, gap length, secondary resistivity, etc. It was also found that the exit-end-effect wave decays so fast that it has little influence on motor characteristics although it distorts the magnetic field near the exit end, and that the entry-end-effect wave decays rather slowly and in the high-speed motor it is present along the entire longitudinal length of the air gap and degrades the performance of the high-speed motor.

(3) *Calculation of Performance under Influence of End Effect*

Our solution made it possible to calculate separately thrusts produced by a normal wave, an entry-end-effect wave and an exit-end-effect wave. Thus the influence of the end-effect wave on thrust was made clear. Our solutions also made it possible to calculate other performance parameters, such as current, power factor, efficiency, etc., for motors under the influence of the end effect and those not under it. This would be the first time that calculated and measured performances were in good agreement for a high speed linear induction motor. The calculations led to the conclusion that the high-speed linear induction motor is capable of outstanding performance if no end effect exists, and that the end effect degrades the performance so much that some means of remedying it is absolutely essential.

(4) *Difference between High-speed Motors and Low-speed Motors with Respect to Normal Performance*

There are some differences in the normal performances of high-speed induction motors and low-speed motors, regardless of whether or not they are rotary types or linear types. In general, the performance of high-speed induction motors is superior to that of low-speed induction motors. This is especially true for a high-speed linear induction motor which has a much higher speed and a longer pole pitch. We also derived formulas for calculating the normal performances of a linear induction motor not under the influence of the end effect, and made it possible to determine influences of parameters, such as gap length, power supply frequency, synchronous speed, pole pitch, secondary resistivity, etc., on motor performance. It was found that if other parameters, such as pole pitch and power supply frequency, are selected properly, the high-speed linear induction motor is capable of very outstanding performance showing very high power factor and very high efficiency, in spite of the very large lateral length of the air gap, as large as about 60 mm or longer. An important fact revealed by our research was that the end effect destroys the outstanding performance of the high-speed linear induction motor, which could be developed if there were no end effect. This fact has not been taken into account in much of the research and experiments on high-speed linear induction motors which propose and recommend rather complicated secondary structures, such as composite secondary or sandwiched structures, both of which have iron parts and which are intended to reduce reluctance in a longer air gap. These complicated secondary structures are not necessary and involve excessive attractive force between the primary and the secondary.

(5) *End-Effect Influence*

Our research revealed that influences of the end effect in high-speed and low-speed linear induction motors are quite different. The most conspicuous differences are as follows; In low-speed motors, the speed of the end-effect wave can be higher than the motor speed and even much higher than the synchronous speed, while in high-speed motors the speed of the end-effect wave is about the same as the motor speed and cannot be higher than the synchronous speed. In low-speed motors, the attenuation of the entry end-effect wave is quick, while in high-speed motors the attenuation is very slow and the entry-end-effect wave is present over the entire longitudinal length of the air gap. As a consequence of the differences, the influence of the end-effect wave on motor performance is also quite different in high-speed motors and low-speed motors. In low-speed motors, the end-effect wave improves motor performance in the low-slip region, the important motor-run region, increasing thrust, power factor and efficiency, and allowing net thrust to be generated even at synchronous and higher speeds. On the contrary, in high-speed motors thrust, power factor and efficiency are reduced to a large extent in the low-slip region, and it is not an overstatement to say that high-speed applications of linear induction motors may not be feasible if the end effect is overlooked and is allowed to remain as an influence. The differences between high-speed and low-speed linear induction motors have not been recognized, and the results and conclusions obtained from research on low-speed linear induction motors have been applied to high-speed linear induction motors, and as a result serious end-effect influences have been overlooked.

(6) *Alleviation of the End Effect*
Our theory revealed a detailed structure of the end effect and made it possible to determine parametric influence of various factors on the normal performance and the end effect of the linear induction motor. Since the one-dimensional theory gives the solutions of the air gap field in the form of explicit functions of the motor parameters, it is very easy to determine from the theory general tendencies of the parametric influence of the important factors, such as number of poles, air gap length, secondary resistivity, supply frequency, etc. Adjustment of these factors in the direction of alleviating the end effect tends to degrade the normal motor performance, making it necesarry to find a compromise. Our theory provided means of finding a good compromise in cases where such a compromise is possible and sufficient for purposes under consideration.

(7) *End-effect Compensation*
It was clearly established above that the high-speed linear induction motor is capable of excellent performance in spite of its large air gap, if the end-effect influences are removed. Feasibility of developing practical high-speed

linear induction motors may be doubtful if the influence of the end effect on motor performance is not removed. The prevention of this influence is an absolute necessity for good motor performance.

We contrived two types of linear induction motors for this purpose; one is the compensated linear induction motor and the other is the linear induction motor of the wound-secondary type. The theory of compensating the end effect was derived from our two-dimensional solution of the air gap field. The compensation theory led to the development of various types of compensated linear induction motors. The most economical type among them was that with a compensating winding with two poles placed in front of the entry end of the main winding. Calculations on the performance of the compensated linear induction motor were made to show that the compensating winding considerably improves high-speed linear induction motor performance which has been degraded by the end effect.

Since the wound secondary carries a uniform ampere conductor over the entire longitudinal length of the air gap, the type of end effect mentioned above is eliminated entirely. However, our theory revealed that an entirely different kind of end effect occurs and influences the performance of a linear induction motor of the wound-secondary type to a large extent. We found a way to eliminate this end effect and applied it to the linear induction motor of the wound-secondary type without iron parts. Calculations on the performance of the linear induction motor were made to show that the performance was ideally suited for the traction of high-speed-trains. This suggested that the linear induction motor of the wound-secondary type provides a possibility of developing an entirely new type of electric railway.

Chapter 2

Type and Model of Linear Induction Motor Used for Investigation

2.1. TYPE OF LINEAR INDUCTION MOTOR USED FOR INVESTIGATION

There are many conceivable types of linear induction motors; however, for high-speed vehicle propulsion applications the number of feasible types is rather limited. Of these, the most suitable may be the two-sided primary, doubly-excited, short-primary, long-secondary type, a sketch of which is shown in Fig. 1. It is the two-sided primary type because there are two primary iron cores, one on each side of the secondary. It is a doubly-excited type because there are two primary windings, each of which is installed in slots respectively on each surface of the two primary iron cores. And it is the short-primary, long-secondary type because the secondary is much longer than the primary. The two-sided, doubly-excited type has a larger output than the one-sided type, and the attractive or repulsive force between the primary and secondary sides is much smaller than in the one-sided type. The short-primary type is much less expensive than the long-primary type because construction of the primary side is much more complicated than that of the secondary. We shall confine our discussion to the type of motor described above. This type is also called the vertical type, because it is usually installed vertically. It should be pointed out here that the same analysis is applicable without modification to both double-sided and single-sided linear induction motors.

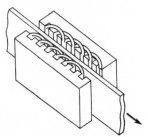

Fig. 1. Two-sided primary, doubly-excited, short-primary, long-secondary type of linear induction motor.

The problem in regard to which side of the motor should be movable is of great importance from the practical point of view. The primary-on-vehicle, secondary-on-ground type is economically advantageous for train propulsion because the number of primary sides needed is much less. There is, however, a technical drawback in that the power supply, which is connected to the primary, must be located on the vehicle. The primary-on-ground, secondary-on-vehicle type is economically disadvantageous because of the much greater number of primary sides installed along the railway. There is, however, a great technical advantage in that the power supply can be directly connected to the primary, eliminating the need for current collection through a trolley.

From the theoretical point of view, it does not matter which of the two sides is installed on the ground. What matters is the relative speed between the primary and the secondary. In the following discussions it will be assumed that the primary is fixed and the secondary moves in the air gap between the two primary cores, as shown in Fig. 1.

2.2. Primary Winding[3]

The primary winding of the linear induction motor is essentially the same as that of the rotating induction motor. When balanced three-phase currents flow in the winding, they must produce a spacially sinuosidal, travelling, magnetomotive force. The primary winding of the linear induction motor is obtained by cutting the primary iron core of the rotating induction motor in a plane through the center line of the shaft and flattening it out. When the primary winding of the rotating motor is a single-layer winding, the corresponding winding of the linear induction motor is also a single-layer winding as shown in Fig. 2. Flux patterns of the linear induction motor with a single-layer winding are shown in Fig. 3(a) and (b). Corresponding flux density distributions in the air gap are shown, respectively, in Fig. 4(a) and (b), which are $\pi/2$ electrical radians apart in phase from each other. The total flux in the core in Fig. 3(b) is twice that of Fig. 3(a). In the rotating motor the flux pattern remains always the same as that of Fig. 3(a). The thickness of the core of the linear induction

Fig. 2. Single-layer winding.

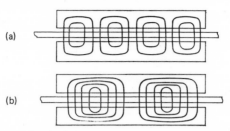

(a)

(b)

Fig. 3. Flux patterns of the single-layer winding shown in Fig. 2 and the double-layer winding shown in Fig. 5(b).

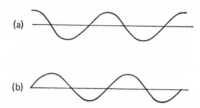

(a)

(b)

Fig. 4. Flux density distribution in the air gap.

motor must then be increased to twice the thickness of that of a corresponding rotating motor in order to provide a magnetic path for the larger amount of core flux in Fig. 3(b). As stated in several articles[1,20] this appears to be a drawback of the single-layer winding.

However, as will be explained later, the flux distribution in the air gap is so distorted by the end effect that flux patterns in Fig. 3 do not actually exist. The end effect weakens the magnetic field near the entry end of the air gap and the magnetic field builds up rather gradually along the longitudinal length of the air gap. Then the above-mentioned drawback of the single-layer winding does not actually exist.

When the double-layer winding is adopted, the problem of arranging the conductors near both ends of the air gap arises. When the primary iron core of the rotating motor is cut and flattened out, the end connections of the primary coils, which cross the intersection plane, are also cut. If the severed coils are omitted from the resulting linear induction motor, the primary winding shown in Fig. 5(a) is derived. This is a double-layer winding, except for the first and last pole pitches, where the winding is of the single-layer type. Flux patterns for this winding are shown in Fig. 6(a) and (b), corresponding, respectively, to the flux distributions in the air gap shown in Fig. 3(a) and (b). The two flux patterns in Fig. 6 are $\pi/2$ apart in phase and have the same maximum flux in the core. Therefore,

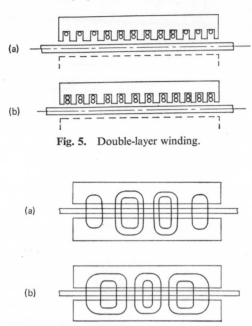

(a)

(b)

Fig. 5. Double-layer winding.

(a)

(b)

Fig. 6. Flux pattern of the double-layer winding shown in Fig. 5(a).

core thickness can be the same as that of the rotating motor and this is claimed by some researchers[1,20] to be an advantage of this winding. Amperes per meter in the first and last pole pitches is half that of the other pole pitches. The same people maintain that building up of the current in two steps at the entry end of the air gap may reduce the end effect.[20] We will show in later chapters that it is not true. On the contrary, the entry-end effect occurs at two steps; the first occurring at the entry end and the second at a distance of one pole pitch from the entry end. Thus the influence of the end effect is prolonged and increased. When the end effect exists along most of the longitudinal length of the air gap, the flux pattern shown in Fig. 6 is not actually present, and the advantage of this winding, as claimed by some researchers, is lost.

It is possible to have two layers of coil sides in the all slots as shown in Fig. 5(b). This can be done in several ways, as shown in Fig. 7 for a two-pole linear induction motor. Current distribution in all the slots is the same in all three windings in Fig. 7, except at the two ends. Since the flux path closes only through the air gap, flux cannot be produced by current in the coil sides at the ends.* Therefore, all three windings produce the same flux distribution in the air gap. The winding in Fig. 7(a) consists of the same type of coils and has an advantage over the other windings.

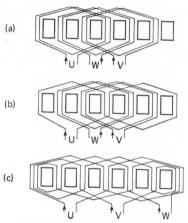

Fig. 7. Double-layer windings.

The single-layer winding in Fig. 2 and the double-layer winding in Fig. 5(b) have the same current distribution over the entire surface of the iron core facing the air gap and produce the same flux patterns. In these windings, the end effect at the entry end of the air gap occurs in one step. It then becomes easier to compensate the end effect, as will be explained in Chapter 13. The amount of the total flux in the core of the compensated linear induction motor, as will be explained in Chapter 13, does not increase to twice that of the corresponding rotating induction motor. Based on these considerations, the single-layer winding shown in Fig. 2 or the double-layer winding shown in Fig. 5(b) is considered superior to the double-layer winding in Fig. 5(a), where the current density in the first and last poles is half of that in other poles and they are not so active in thrust production.

Winding connections are classified into two kinds; series connection and parallel connection. In series connection all phase belts of each phase are connected in series and there is only one circuit within each phase. In parallel connection parallel circuits are used; the largest number of parallel circuits is equal to the number of poles. In rotating motors there is no difference between performances of the two winding connections, except

---

* Leakage flux, which escapes from the iron core at the side or back, is very slight and negligible.

that with the same total number of conductors, the series connection provides higher voltage and less current and the parallel connection provides lower voltage and more current. This situation also exists in the linear induction motor when it is under constant current drive. The situation changes, however, when it is under constant voltage drive. In later chapters, we will show how the end effect in the linear induction motor weakens the magnetic field near the entry end of the air gap. Accordingly, in the parallel-connected motor, phase belts of the winding located nearer the entry end draw more current from the constant voltage source than phase belts located farther from the entry end, resulting in a nonuniform electromotive force of the primary winding and excess copper loss in phase belts located nearer the entry end. This is an undesirable situation which makes analysis of the air-gap field more difficult. In the series-connected motor, same current flows from the entry end to the exit end, resulting in a uniform electromotive force in the primary winding. In the analyses and discussions that follow all motors are series connected motors unless stated otherwise.

2.3. EQUIVALENT CURRENT SHEET AND A MODEL
     OF THE LINEAR INDUCTION MOTOR

The induction motor has, in most cases, three-phase primary winding with the conductors inserted in slots on both surfaces of the primary iron cores, as shown in Fig. 1. As explained in Section 2.2, it is assumed that all the slots are filled with an equal number of conductors and all the coils of each phase are connected in series. The same current then flows from the entry end to the exit end of each phase and equal amount of ampere-conductors exists in all the slots of each phase.

In field theories of the electrical machinery, currents in windings are usually represented by current sheets. The actual current in a winding is a discrete quantity concentrated in conductors located in slots, while a current sheet is a continuously distributed quantity. However, they are practically equivalent to each other, if both produce the same sinusoidal magnetomotive force as the fundamental components of their Fourier expansion.

The primary current i_1 is represented by

$$i_1 = \sqrt{2}\, I_1 \exp(j\omega t). \tag{1}$$

Currents in other phases of the primary windings lag by $2\pi/3$ and $4\pi/3$, respectively. Combined action of the three-phase currents of the primary winding produces a travelling magnetomotive force given by the following formula

$$A_1 = \frac{3\sqrt{2}}{\pi} \frac{w_1 k_{w_1} I_1}{p_1} \exp\left\{ j\left(\omega t - kx - \frac{\pi}{2}\right)\right\}. \tag{2}$$

Here k is related to the pole pitch τ by the following formula

$$k = \frac{\pi}{\tau}. \tag{3}$$

It is assumed that the equivalent current sheet has the following current distribution.

$$j_1 = J_1 \exp\{j(\omega t - kx)\} \qquad (\text{A/m}). \tag{4}$$

Here k is given by Eq. (3) which means that both the primary winding and its equivalent current sheet have the same pole pitch, as they should. j_1 of Eq. (4) produces the following magnetomotive force

$$A_1 = \frac{J_1}{k} \exp\left\{ j\left(\omega t - kx - \frac{\pi}{2}\right)\right\}. \tag{5}$$

This must be identical to Eq. (2), hence we obtain

$$J_1 = \frac{3\sqrt{2}}{\pi} \frac{k w_1 k_{w_1} I_1}{p_1}$$

$$= \frac{3\sqrt{2}}{p_1\tau} w_1 k_{w_1} I_1 \qquad (\text{A/m}). \tag{6}$$

The current sheet thus determined by Eqs. (4) and (6) is the equivalent current sheet, which produces the same sinusoidal magnetomotive force as that of the primary winding.

The secondary side of the linear induction motor is usually made of a metallic sheet which is in most cases a homogenious, conductive, sheet. However, in some instances, it is made of several kinds of metallic sheets; such as an iron sheet covered on both surfaces with copper or aluminum sheets, or it can also be made of copper (or aluminum) bars interleaved with laminated iron in the same way as in the squirrel cage induction motor. These composite secondaries have rather complicated construction and are accompanied by higher cost. They also involve the problem of lateral attractive magnetic force, which makes maintenance of the air gap difficult, because of the presence of iron on the secondary. In the following discussions, it will be assumed, unless otherwise stated, that the secondary side is a homogeneous, conductive nonmagnetic sheet.

As explained above, the linear induction motor is represented by primary iron cores with current sheets on their surfaces facing the air gap, and a secondary conductive sheet. Linear induction motors of the short-primary, long-secondary type and the two-sided primary type are represented in the following theoretical analyses by the model shown in Fig. 8. The model has primary current sheets on both surfaces of the primary iron cores. The current distribution of the primary current sheets is given by Eqs. (4) and (6). The model is an equivalent linear induction motor under constant current drive, where the primary current is given. One omission in the model is the space harmonics of the primary magnetomotive forces. This omission usually occurs in ordinary theories of the induction motor, and in order to compensate for it, the space harmonics are included in the primary leakage inductance. This compensation is also applicable to the model and appropriate primary leakage impedance should be inserted in series with the electromotive force induced by the sinusoidal travelling magnetic field in the primary winding. This way of calculating the terminal voltage of the primary winding will be explained in Chapter 6. The model is then also applicable to the linear induction motor under constant voltage drive.

Chapter 3

Equations for the Electromagnetic Field in the Air Gap and Their Solutions Based on a One-dimensional Model*

3.1. Field Equations and General Solutions

In the model of the linear induction motor shown in Fig. 8, the primary current is represented by the two current sheets whose thickness is infinitesimally small. The coordinate axes are chosen as shown in the figure. The primary current flows in z direction and travels in the direction of the motion of the secondary sheet, which is also the direction of the x coordinate. The secondary conductive sheet has finite thickness. It is probable that the variation of the field within the gap is two dimensional, that is, variation in the x coordinate and in the y coordinate, which is perpendicular to the x coordinate. If the secondary conductive sheet is non-magnetic, variation in the y coordinate is small, as will be shown in Section

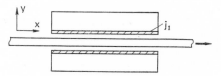

Fig. 8. Model of a linear induction motor with primary windings represented by current sheets.

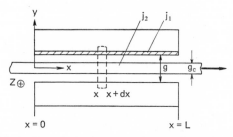

Fig. 9. Model of a linear induction motor for one-dimensional analysis.

* Reported originally in refs. 4, 9 and 13.

10.3, and can be neglected for many practical purposes. Then the one-dimensional model, which takes only the variation in the x coordinate into consideration, is sufficient for many purposes. The one-dimensional model in this sense is reproduced in Fig. 9. The origin of the x coordinate is located at the entry end; and the x coordinate at the exit end is L. The primary current j_1 in the current sheet and the secondary current j_2 flow in the z direction. In the figure, j_1 flows on one side of the primary cores, and is the sum of the primary currents on both sides. The thickness of the iron stack in the z direction is assumed to be 1 m. The line integral of the magnetic field H around the periphery of a rectangle located at x and Δx wide, is $g(\partial H/\partial x)\Delta x$. The amount of current enclosed by the periphery of the rectangle is $(j_1+j_2)\Delta x$. The two quantities are equal according to the Ampere's law and we obtain

$$g\frac{\partial H}{\partial x}=j_1+j_2 . \tag{7}$$

Multiplying it by the permeability of air

$$\frac{\partial b}{\partial x}=\frac{\mu_0}{g}(j_1+j_2). \tag{8}$$

The electromotive force e_2 in the secondary is directed in the z coordinate. The line integral of e around the periphery of a rectangle, 1-m long and Δx wide, placed on the secondary sheet located at x is $(\partial e/\partial x)\Delta x$. The electromotive force has two components; the transformer voltage which contributes $(\partial b/\partial t)\Delta x$ and the speed voltage which contributes $v(\partial b/\partial x)\Delta x$, where v is the speed of the secondary sheet. Equating the line integral of the electromotive force to the sum of the transformer voltage and the speed voltage, we obtain

$$\frac{\partial e_2}{\partial x}=\frac{\partial b}{\partial t}+v\frac{\partial b}{\partial x}. \tag{9}$$

In the present one-dimensional model, flux density b is directed in the y coordinate and has no x component. This means that the secondary sheet has no leakage inductance. Therefore the secondary electromotive force e_2 is completely consumed as resistance drop. Thus we have

$$e_2=j_2\rho_{\mathrm{s}} . \tag{10}$$

Here ρ_{s} is the surface resistivity of the secondary sheet and is given by

$$\rho_{\mathrm{s}}=\frac{\rho}{g_c} \quad (\Omega), \tag{11}$$

where ρ is the volume resistivity of the secondary sheet, g_c its thickness. From Eqs. (8), (9) and (10) the following equation is obtained:

$$\frac{g}{\mu_0}\frac{\partial^2 b}{\partial x^2} - \frac{v}{\rho s}\frac{\partial b}{\partial x} - \frac{1}{\rho s}\frac{\partial b}{\partial t} = \frac{\partial j_1}{\partial x}. \tag{12}$$

We shall now seek solutions for this equation in the case of the primary current sheet being excited by the following current:

$$j_1 = J_1 \exp\left\{ j\left(\omega t - \frac{\pi}{\tau}x\right)\right\}$$

$$= J_1 \exp\left\{ j\frac{\pi}{\tau}(v_s t - x)\right\} \quad \text{(A/m)}. \tag{13}$$

This equation is essentially the same as Eq. (4), and here τ is the pole pitch of the primary winding, ω the angular frequency of the power supply, and v_s the synchronous speed.

The steady state solution should be a travelling sinusoidal wave at synchronous speed with the following form:

$$b_s = B_s \exp\left\{ j\left(\frac{\pi}{\tau}v_s t - \frac{\pi}{\tau}x + \delta_s\right)\right\}. \tag{14}$$

By inserting Eqs. (13) and (14) into Eq. (12), we have

$$B_s = \frac{J_1}{\sqrt{\left(\frac{\pi g}{\tau \mu_0}\right)^2 + \left\{\frac{1}{\rho s}(v_s - v)\right\}^2}}, \tag{15}$$

$$\delta_s = \tan^{-1}\frac{\pi \rho s g}{\mu_0 \tau(v_s - v)}. \tag{16}$$

Nonsteady solutions are obtained from the following equation, which is obtained by setting the left side of Eq. (12) equal to zero.

$$\frac{g}{\mu_0}\frac{\partial^2 b}{\partial x^2} - \frac{v}{\rho s}\frac{\partial b}{\partial x} - \frac{1}{\rho s}\frac{\partial b}{\partial t} = 0. \tag{17}$$

It is assumed that the solution of Eq. (17) has the following form of separated variables:

$$b(t, x) = T(t)B(x). \tag{18}$$

Inserting Eq. (18) into Eq. (17) and introducing an arbitrary constant λ for separating variables t and x from each other, we get

$$\frac{1}{T(t)}\frac{dT(t)}{dt} = \frac{1}{B(x)}\left\{\frac{\rho s g}{\mu_0}\frac{d^2 B(x)}{dx^2} - v\frac{dB(x)}{dx}\right\}$$

$$\equiv \lambda. \tag{19}$$

Taking $T(t)$ out of Eq. (19), we have

$$\frac{dT(t)}{dt} = \lambda T(t), \tag{20}$$

whose solution is

$$T(t) = C_\varepsilon^{\lambda t}, \tag{21}$$

where C is an arbitrary constant.

$B(x)$ part of Eq. (19) is as follows:

$$\frac{dB(x)}{dx^2} - \frac{\mu_0 v}{\rho s g} \frac{dB(x)}{dx} - \frac{\mu_0 \lambda}{\rho s g} B(x) = 0. \tag{22}$$

Assuming $B(x)$ has the form B_ε^{kx}, and inserting it into Eq. (22), we get

$$k^2 - \frac{\mu_0 v}{\rho s g} k - \frac{\mu_0 \lambda}{\rho s g} = 0, \tag{23}$$

whose solutions are

$$k_1, k_2 = \frac{\mu_0 v}{2 \rho s g} \pm \frac{1}{2} \sqrt{\left(\frac{\mu_0 v}{\rho s g}\right)^2 + \frac{4 \mu_0 \lambda}{\rho s g}}. \tag{24}$$

Constant λ in Eq. (19) is arbitrary and may be real, imaginary or complex. Given boundary conditions of problems under consideration prescribe a set of λ and let it be denoted by λ_n. Then Eq. (18) now can be written as

$$b(t, x) = \exp(\lambda_n t) \{B_{n1} \exp(k_{n1} x) + B_{n2} \exp(k_{n2} x)\}. \tag{25}$$

Here k_{n1} and k_{n2} are k_1 and k_2 of Eq. (24), which correspond to λ_n, and B_{n1} and B_{n2} are arbitrary constants. The general solution of Eq. (12) is obtained by combining b_s of Eq. (14) and $b(t,x)$ of Eq. (25) and can be expressed as follows:

$$b(t, x) = B_\mathrm{s} \exp\left\{j\left(\omega t - \frac{\pi}{\tau} x\right)\right\}$$
$$+ \sum_n \exp(\lambda_n t) \{B_{n1} \exp(k_{n1} x) + B_{n2} \exp(k_{n2} x)\}. \tag{26}$$

3.2. Solution for Steady Operation

Equation (26) is a general solution for the magnetic flux density in the air gap of the linear induction motor and is applicable to both a transient condition and steady operation of the linear induction motor. λ_n in Eq. (26) must be chosen properly so that the equation can satisfy boundary conditions of a given problem. For steady operation of the motor, all the terms of Eq. (26) must be steady. This prescribes that λ_n must be pure imaginary and must not contain a real part. Since a sole exciting source is the primary current and it has the angular frequency ω, λ_n can take $j\omega$ only.

$$\lambda_n = \lambda = j\omega. \tag{27}$$

Inserting Eq. (27) into Eq. (24), we get

$$k_1, k_2 = \frac{\mu_0 v}{2\rho s g} \pm \frac{1}{2}\sqrt{\left(\frac{\mu_0 v}{\rho s g}\right)^2 + 4j\frac{\omega\mu_0}{\rho s g}}. \tag{28}$$

The following equation is introduced.

$$\sqrt{\left(\frac{\mu_0 v}{\rho s g}\right)^2 + 4j\frac{\omega\mu_0}{\rho s g}} = X + jY, \tag{29}$$

where X and Y are obviously positive real. From Eqs. (28) and (29), we have

$$k_1 = \frac{\mu_0 v - \rho s g X}{2\rho s g} - j\frac{Y}{2} = -\frac{1}{\alpha_1} - j\frac{\pi}{\tau_e}, \tag{30}$$

$$k_2 = \frac{\mu_0 v + \rho s g X}{2\rho s g} + j\frac{Y}{2} = \frac{1}{\alpha_2} + j\frac{\pi}{\tau_e}, \tag{31}$$

whence the following quantities are obtained.

$$\alpha_1 = \frac{2\rho s g}{\rho s g X - \mu_0 v}, \tag{32}$$

$$\alpha_2 = \frac{2\rho s g}{\rho s g X + \mu_0 v}, \tag{33}$$

$$\tau_e = \frac{2\pi}{Y}. \tag{34}$$

Inserting Eqs. (30)–(34) into Eq. (26), the solution for steady operation of the linear induction motor is given by

$$b = B_s \exp j\left(\omega t - \frac{\pi}{\tau}x\right)$$

$$+ B_1 \exp\left(-\frac{x}{\alpha_1}\right)\exp j\left(\omega t - \frac{\pi}{\tau_e}x\right)$$

$$+ B_2 \exp\left(\frac{x}{\alpha_2}\right)\exp j\left(\omega t + \frac{\pi}{\tau_e}x\right). \tag{35}$$

All the three terms of this equation are steady with respect to time t. The first term is the normal travelling wave given by Eq. (14); B_s corresponding to $B_S \exp(j\delta_S)$. B_1 and B_2 are arbitrary functions of t to be determined from boundary conditions. α_1, α_2 and τ_e are all positive, as can be seen from Eqs. (32)–(34). The second term of Eq. (35) is an attenuating travelling wave which travels in the positive direction of x and whose attenuation constant is $1/\alpha_1$ and half-wave length is τ_e. The third term of Eq. (35) is an attenuating travelling wave, which travels in the negative direction and whose attenuation constant is $1/\alpha_2$ and half-wave length is also τ_e. The B_1 wave is caused by the core discontinuity at the entry end and the B_2 wave is caused by the core discontinuity at the exit end, hence, both are called "end-effect waves." Both end-effect waves have an angular frequency ω, which is the same as that of the power supply. They have the same half-wave length τ_e, which is different from half-wave length τ (equal to pole pitch) of the primary winding. The travelling speed of the end-effect waves is given by

$$v_e = \frac{\omega \tau_e}{\pi} = v_S \frac{\tau_e}{\tau}$$

$$= 2f\tau_e . \tag{36}$$

Since τ_e is a function of the speed of the secondary sheet through Eq. (34), the speed v_e of the end effect wave is also a function of v.

Equation (29) can be approximated, when the real part is much larger than the imaginary part, as follows

$$\sqrt{\left(\frac{\mu_0 v}{\rho_S g}\right)^2 + 4j\frac{\omega\mu_0}{\rho_S g}} \doteq \frac{\mu_0 v}{\rho_S g} - j\frac{2\omega}{v} \quad \text{for} \quad \frac{\mu_0 v^2}{4\omega\rho_S g} \gg 1. \tag{37}$$

Comparing Eq. (29) and Eq. (37) we get

$$\left. \begin{array}{l} X \doteq \dfrac{\mu_0 v}{\rho_S g}, \\[3mm] Y \doteq \dfrac{2\omega}{v} = \dfrac{4\pi f}{v} . \end{array} \right\} \tag{38}$$

Inserting Y of Eq. (38) into Eq. (34), we have

$$\tau_e \doteq \frac{v}{2f} = \tau(1-s), \tag{39}$$

where s is the slip. Insertion of Eq. (38) into Eq. (35) gives

$$v_e \doteq v_S(1-s) = v. \tag{40}$$

This equation indicates that the speed of the end-effect waves is the same as that of the secondary sheet, if the inequality condition in Eq. (37) holds, that is, if the motor speed is high. However, it should be mentioned that v_e can be much higher than v in low-speed motors, as will be shown later.

In Eq. (35), B_S, B_1 and B_2 are complex variables of t. B_S is equal to $B_S \exp(j\delta_S)$ in Eq. (14), and B_1 and B_2 are determined from boundary conditions. Two boundary conditions are necessary to determine B_1 and B_2. One condition is derived from the fact that the total amount of flux leaving the primary iron core is zero. If flux does not leave the iron core from the back and sides and there is no fringing at both ends of the air gap, then flux exists only in the air gap and a mathematical expression for the above-mentioned fact is given by the following formula:

$$\int_0^L b\,dx = 0, \tag{41}$$

where L is the core length in the direction of the x coordinate. Another boundary condition is given in Appendix III, in which Eq. (A-22) gives

$$b\Big|_{x=0} = -\frac{\rho_S}{v} j_1 \Big|_{x=0}. \tag{42}$$

In Eq. (35), B_S term is the normal forward wave, and B_1 term is the forward end-effect wave generated at the entry end and will hereafter be called the entry-end-effect wave. B_2 term is the backward wave generated at the exit end and will hereafter be called the exit-end-effect wave. The entry-end wave attenuates while travelling in the positive direction of the x coordinate. The length of penetration of the entry-end-effect wave is α_1. The exit-end wave attenuates while travelling in the negative direction of the x coordinate. The length of penetration of the exit-end-effect wave is α_2. As can be seen from Eqs. (32) and (33), and as later numerical examples will show, $\alpha_1 \gg \alpha_2$. The entry-end-effect wave may reach the exit end of the air gap, when the length of penetration α_1 is not much shorter than the core length L. And this is the case in most practical examples, especially in high-speed motors. On the other hand the length of penetration α_2 is very short and is always much shorter than the core length L. Under such a condition, the exit-end-effect wave B_2, which is a backward travelling wave, can not reach the entry end. The B_1 wave covers the entire core length and its influences on the air-gap magnetic field and motor performances are considerable. On the other hand, the B_2 wave can exist only in the vicinity of the exit end and its influences are very small.

At very high speed Eq. (42) becomes

$$b\big|_{x=0} = 0. \tag{43}$$

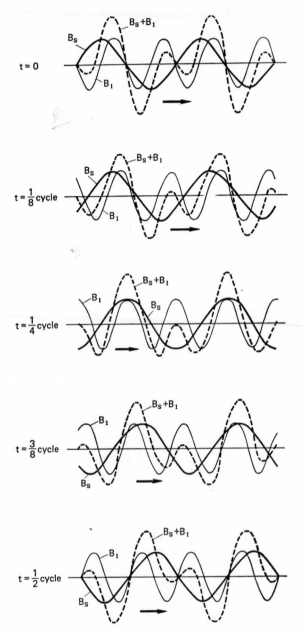

Fig. 10. Normal wave (B_S), entry-end-effect wave (B_1) and resultant wave (B_S+B_1).

Near the entry end the existence of the exit-end-effect wave (B_2 wave) can be neglected. Then Eq. (43) gives

$$B_1 = -B_S \qquad \text{for} \quad x = 0. \tag{44}$$

This means that the entry-end-effect wave is equal in amplitude and opposite in phase to the normal wave at the entry end of the air gap, and that the former almost cancels the latter in the vicinity of the entry end. Since the speed v_e of the B_1 wave given by Eq. (36) is different from that of B_S wave (synchronous speed v_S), the relative phase difference between both waves varies as they travel in the gap. When α_1 is much larger than L and $v_e = 0.5v_S$, the behavior of the B_S wave and B_1 wave in the air gap is as shown in Fig. 10. The resultant magnetic flux density is almost zero at the entry end and its maximum value can reach twice the amplitude of the normal wave. In high-speed motors, both the B_S wave and B_1 wave have the same speed when slip is zero, as shown by Eq. (40). Their relative phase difference then remains the same over the entire length of the iron core, and they cancel out each other if α_1 is much longer than the core length. In this state, the air gap field becomes very weak over the entire length of the iron core, and it is anticipated that motor performance becomes very poor near the synchronous speed.

Chapter 4

Results of Calculation on the Magnetic Field in the Air Gap*

In this chapter some examples of the calculated results of analytic solutions of the field equations for the air gap of the one-dimensional model in Fig. 9 will be given.

Length of penetration α_1 and α_2 are given by Eqs. (32) and (33), respectively. They are measures of the distances to which end-effect waves,

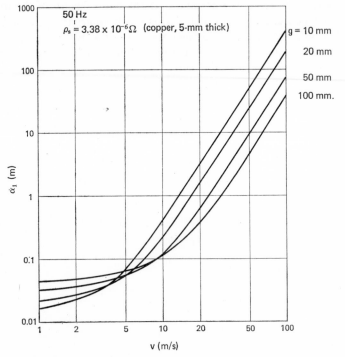

Fig. 11. Calculated values of length of penetration α_1 of entry-end-effect wave versus motor speed v.

* Reported originally in ref. 13.

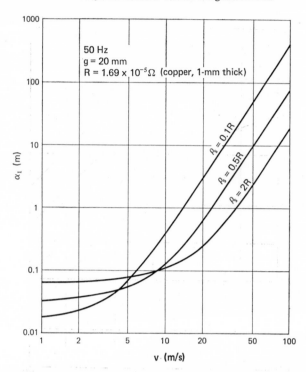

Fig. 12. Calculated values of length of penetration α_1 of entry-end-effect wave versus motor v.

B_1 or B_2, can penetrate into the air gap from the entry end or exit end. Figures 11, 12, 13 and 14 are the calculated results for cases where the power supply frequency is 50 Hz. In Fig. 11 surface resistivity ρ_s of the secondary sheet is 3.38×10^{-6} Ω, which corresponds to 5-mm-thick copper sheet. The gap length g^* is indicated for each curve. α_1 increases as speed v of the secondary sheet increases. For v around 100 m/s, α_1 is very large; for example, $\alpha_1 \doteqdot 200$ m for $g = 20$ mm. This value of α_1 is much larger than most primary core lengths, and in such cases the entry-end-effect wave prevails over the whole length of the primary core. In Fig. 12, secondary resistivity ρ_s is given as a parameter for each curve. It should be noted that in Figs. 11 and 12 the natures of the α_1-v characteristic curves are quite different in the lower-speed region and in the higher-speed region. In the

* Gap length includes thickness of the nonmagnetic secondary sheet.

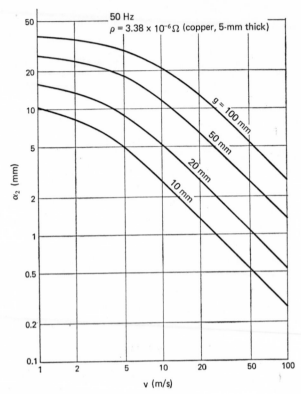

Fig. 13. Calculated values of length of penetration α_2 of exit-end-effect wave versus motor speed v.

higher-speed region, the curves are straight and steep, compared with the lower-speed region, where curves are almost horizontal. The parametric influences of gap length g and secondary surface resistivity ρ_s are also quite different in the two regions. In the higher-speed region of Fig. 11, the larger g is, the smaller α_1 becomes, while in the lower-speed region it is the opposite. In the higher-speed region of Fig. 12, the larger ρ_s is, the smaller α_1 becomes, while in the lower-speed region it is the opposite. As will be explained later, high-speed linear induction motors behave quite differently from low-speed linear induction motors. The criterion for separating low-speed linear induction motors and high-speed linear induction motors is given by the following inequality which appeared in Eq. (37),

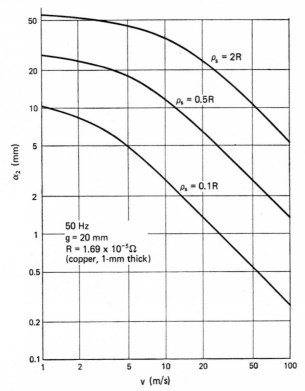

Fig. 14. Calculated values of length of penetration α_2 of exit-end-effect wave versus motor speed v.

$$\frac{\mu_0 v^2}{4\omega \rho_s g} \gg 1. \tag{45}$$

For Figs. 11 and 12, the values of the formula in Eq. (45) are as follows:

$$\frac{\mu_0 v^2}{4\omega \rho_s g} = \frac{4\pi \times 10^{-7} \times 10^2}{4 \times 2\pi \times 50 \times 3.38 \times 10^{-6} \times 0.02}$$

$$\doteqdot 1.5,$$

where

$$\omega = 2\pi \times 50, \quad v = 10 \text{ m/s}, \quad \rho_s = 3.38 \times 10^{-6} \ \Omega \quad \text{and} \quad g = 0.02 \text{ m}.$$

The inequality in Eq. (45) is the condition for high-speed linear induction motors.

Figures 13 and 14 show length of penetration α_2 of the exit-end-effect wave. Compared with α_1, α_2 is much smaller; for example, α_2 is 1 mm for $g = 50$ mm and $v = 100$ m/s. α_2 decreases, as v increases; this tendency is opposite to that of α_1. The exit-end-effect wave is generated at the exit end of the air gap and travels in the direction opposite to that of the normal wave. Since α_2 is very small, the exit-end-effect wave attenuates so fast and its existence is recognizable only in vicinity of the exit end. As will be explained later, the exit-end-effect wave has little influence on motor performance.

The half-wave length τ_e is same for both end-effect waves and is given by Eqs. (29) and (34). The calculated results are shown in Figs. 15 and 16, and clearly a distinction between the low-speed region and the high-speed region exists. In the high-speed region, the half-wave length τ_e is the same irrespective of the parameters indicated in the figures, while in the low-speed region it is influenced by the parameters. In the high-speed region, τ_e is given by Eq. (39) and the speed of the end-effect wave is given by Eq. (40), and the end-effect waves travel at the same speed as the speed of the secondary sheet. This fact indicates that the entry-end-effect wave is

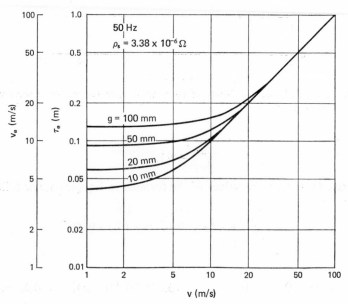

Fig. 15. Calculated values of half-wave length τ_e and speed v_e of end-effect wave versus motor speed v.

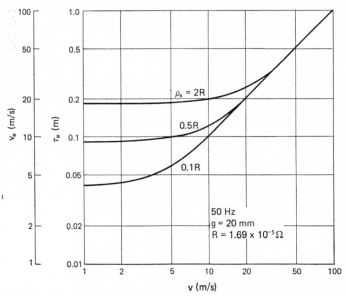

Fig. 16. Calculated values of half-wave length τ_e and speed v_e of end-effect wave versus motor speed v.

carried by the secondary sheet, that is, d-c current flowing in the secondary sheet excites the entry-end-effect wave. This is true for the entry-end-effect wave in high-speed region. In lower-speed region, the half-wave length is longer than that calculated by Eq. (39), and is influenced by parameters, as shown in Figs. 15 and 16. The speed of the end-effect wave is higher than that given by Eq. (40). This means that the end-effect wave travels faster than the secondary sheet in the low-speed region as shown in Fig. 15. It can happen that the entry-end-effect wave travels faster than synchronous speed and can generate considerable thrust at synchronous and higher speeds, as will be shown later.

The magnetic field in the air gap is a combination of the normal wave (B_S wave), the entry-end-effect wave (B_1 wave) and the exit-end-effect wave (B_2 wave). One example of the resultant flux density distribution is shown in Fig. 17. When only the normal wave exists, flux density distribution in the air gap is uniform and flat over the entire length of the primary core. When a linear motor is under constant current drive, the B_S wave is given by Eqs. (14), (15) and (16). Distributions of $|B_S|$ for several slips are shown by broken lines in Fig. 17. Resultant flux density distribution, which

32

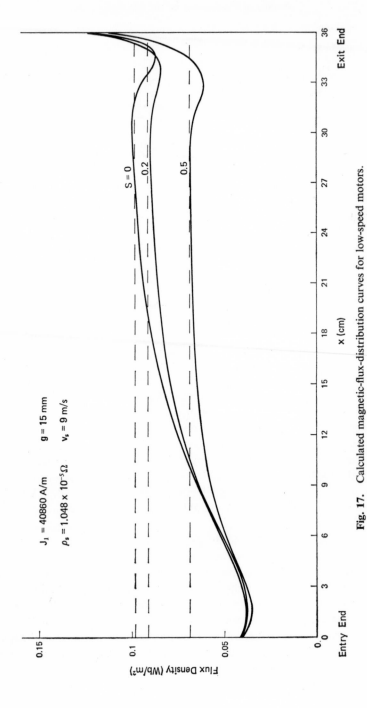

Fig. 17. Calculated magnetic-flux-distribution curves for low-speed motors.

is shown in Fig. 17 for various slips, is given by $|b|$ in Eq. (35). Parameters of the motor are given in the figure, and boundary conditions for the calculation were obtained from Eqs. (41) and (42). It is noted that the field is weakened considerably at the entry end and considerably strengthened at the exit end. A low-speed motor, with a synchronous speed of 9 m/s, is used in the example. The B_1 wave weakens the field at the entry end, and its weakening action extends over a considerable distance depending on the slip. The B_2 wave strengthens the field in vicinity of the exit end and creates a dip in the distribution curves at short distance from the exit end.

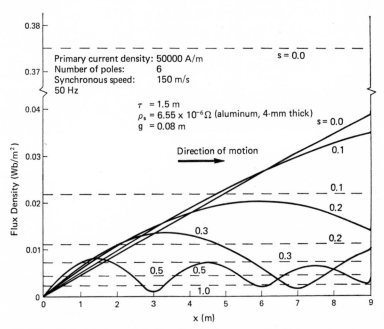

Fig. 18. Calculated magnetic-flux-distribution curves for high-speed motors.

Figure 18 shows the calculated flux distribution curves for a high-speed motor with a synchronous speed of 150 m/s. The curves are different from those in Fig. 17 in that at larger slips, distribution curves fluctuate periodically instead of increasing monotonously. This fluctuation is caused by the B_1 wave whose length of penetration is much longer than the core length.

Chapter 5

Results of Calculation on the Performance of Linear Induction Motor under Constant Current Drive*

As was shown in Chapter 4 the end effect changes the magnetic field in the air gap. It weakens the magnetic field near the entry end, and the weakening effect persists over a much longer portion of the air gap if the motor speed is higher. This influence on the magnetic field naturally extends to motor performance. The influence may be beneficial in some cases and adverse in others. It will be shown later that the beneficial cases are mostly in low-speed motors and adverse cases are in high-speed motors. The performance of high-speed linear induction motors is degraded to such a large extent that feasibility of the linear induction motor might be lost. Some remedies may be necessary to insure the feasibility of the motor. End-effect influences should be taken into account fully in the calculation of linear induction motor performance.

The performance of a linear induction motor under constant current drive is calculated as follows: thrust is given by

$$T_{\text{inst}} = D \int_0^L \text{Re}[j_1] \, \text{Re}[b] dx, \tag{46}$$

where D is the stack length of the iron core, j_1 the primary current given in Eq. (13), and b the magnetic flux density given in Eq. (35). The thrust in Eq. (46) is instantaneous thrust which contains a double-frequency component. The time-average value of the thrust is calculated by

$$T = D \int_0^L \text{Re}[j_1 b^*] dx, \tag{47}$$

where b^* is the conjugate of b in Eq. (35). The methods of calculating power, power factor and efficiency are the same as those used in calculations for motors under constant voltage drive; and power factor and

* Reported originally in refs. 11, 13 and 23.

efficiency are the same for both drives. The method of calculating motor performance under constant voltage drive will be given in Chapter 6. The three linear induction motors listed in the table in Appendix VI will be used as examples for performance calculations.

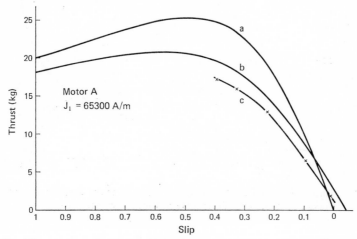

Fig. 19. Thrust-versus-slip characteristics for motor A under constant-current drive. Curve *a*: without end effect; curve *b*: with end effect; curve *c*: measured.

Thrust-versus-slip characteristics was calculated for motor A under a constant current drive of $J_1 = 653$ A/cm, and the result is shown in Fig. 19. Curve *a* shows thrust when there is no end effect and has a typical form for ordinary induction motors, that is, thrust being zero at slip zero. Thrust for this case is given by the following formula,

$$T = D \int_0^L \mathrm{Re}[j_1 b_S{}^*]dx, \tag{48}$$

where $b_S{}^*$ is conjugate of b_S in Eq. (35).

Curve *b* shows thrust when the end effect exists. It is noted that thrust is generated at synchronous speed (9 m/s) and higher speed. In smaller slip region, thrust is increased in relationship with curve *a*. Thrust generation at synchronous speed was confirmed experimentally by curve *c*. This improved characteristic at and near synchronous speed is a special feature of low-speed linear induction motors. The reason for this feature will be explained in Chapter 7.

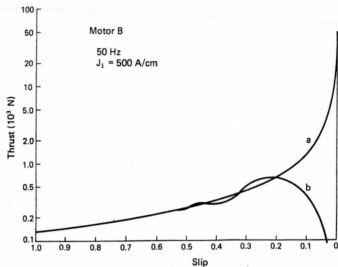

Fig. 20.   Thrust-versus-slip characteristics for motor B under constant current drive. Curve *a*: without end effect; curve *b*: with end effect.

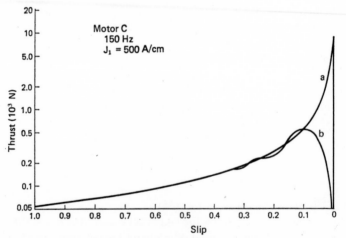

Fig. 21.   Thrust-versus-slip characteristics for motor C under constant current drive. Curve *a*: without end effect; curve *b*: with end effect.

Thrust-versus-slip characteristics was calculated for motors B and C in the table in Appendix VI driven by a constant primary current $J_1 =$

50000 A/m. The synchronous speed of both motors, which are high-speed linear induction motors, is 150 m/s. The results are shown in Figs. 20 and 21. Curve *a* shows thrust when there is no end effect. It was noted that slip for maximum thrust is very small for constant current drive, especially when synchronous speed is high, as indicated in Eq. (70) in Chapter 8. Curve *b* shows thrust when the end effect exists. The influence of the end effect is remarkable in the small-slip region, where thrust is reduced considerably. This is a characteristic peculiar to high-speed linear induction motors, and the reason for this will be explained in Chapter 7. The reduction of thrust is large in the low-slip region, where induction motors usually run and perform best. This phenomenon, therefore, presents obstacles to practical applications and should be fully considered in the development of high-speed linear induction motors.

Chapter 6

Calculations on the Performance of Linear Induction Motor under Constant Voltage Drive*

When a linear induction motor is supplied with a constant voltage source, its performance is different from that under constant current drive. Under constant voltage drive, the primary current changes as slip changes, while it remains the same under constant current drive. In many practical applications, constant voltage drive is usually used, although, in a strict sense, it is rather difficult to realize. Constant current drive is not usually used, although certain types of drive in some applications can be considered, more or less, to be constant current drive within a limited range of slip.

Performance under constant voltage drive is more difficult to analyse because the primary current is not given at the beginning of performance calculations. The method of calculating performance under constant voltage drive will be now explained.

In Chapter 3 magnetic field in the air gap was analyzed based on the one-dimensional model of the linear induction motor. The magnetic flux density distribution, which consists of the normal wave (B_S wave), the entry-end-effect wave (B_1 wave) and the exit-end-effect wave (B_2 wave), was derived as shown in Eq. (35). The derivation was based on the assumption that the primary current was known. After the magnetic field in the air gap is known, it is then possible to calculate the induced electromotive force in the primary winding in the following way.

The number of flux linkages of a coil, whose number of turns is w_C and whose short pitch factor is β, is given by

$$\lambda_C = Dw_C \int_x^{x+\beta\tau} b\,dx, \tag{49}$$

where x and $x+\beta\tau$ are the locations of the coil sides. The total number of flux linkages per phase of the primary winding, which is connected in series, is given by

* Reported originally in refs. 8, 14 and 19.

$$\lambda = \sum_1^P \sum_1^q \lambda_c, \tag{50}$$

where P is number of poles and q is number of coil sides per phase per pole. Since magnetic flux density is in Eq. (35) has the form $b = B(x)B(t) = B(x)\varepsilon^{j\omega t}$, λ_c in Eq. (49) and λ in Eq. (50) have also similar forms and λ in Eq. (50) can be expressed for three phases a, b and c by

$$\lambda_a = \Lambda_a \varepsilon^{j\omega t},$$

$$\lambda_b = \Lambda_b \varepsilon^{j\omega t},$$

$$\lambda_c = \Lambda_c \varepsilon^{j\omega t}. \tag{51}$$

The flux linkages are different in amplitude and phase, since each phase of the three-phase windings is positioned at a different location on the primary iron core with respect to the two ends of the air gap. Accordingly, each of the three phases has a distinctly different induced electromotive force, resulting in a three-phase electromotive force that is markedly unbalanced.

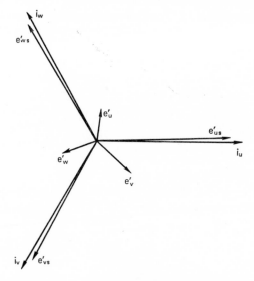

Fig. 22. Phasor diagram of a linear induction motor.

Figure 22 is an example of unbalanced electromotive forces in a three-phase winding due to the end effect. A linear induction motor, which has four poles and is running at a 0.05 slip, has three phase currents i_u, i_v and

i_w, that are assumed to be balanced. When there is no end effect, electromotive forces in the primary winding are induced by the B_S wave in Eq. (35). They are balanced electromotive forces e'_{us}, e'_{vs} and e'_{ws} in Fig. 22. If the end effect is taken into account, the three-phase electromotive forces are induced by three waves shown in Eq. (35), that is, B_S wave, B_1 wave and B_2 wave, and they are indicated by e_u', e_v' and e_w' in Fig. 22. Compared with electromotive forces e_{us}', e_{vs}' and e_{ws}', e_u', e_v' and e_w' are much smaller and unbalanced in amplitude. This indicates that the end effect would considerably influence motor characteristics, as will be shown later.

The unbalanced electromotive forces cause unbalance among the three-phase currents, and degradation in motor performance. The unbalance in the electromotive forces and currents can be removed in the following way. Three linear induction motors, which operate on a common secondary, are grouped together and three primary windings are connected as shown in Fig. 23. Three of the same phase windings are connected in series so that they occupy different positions in the cores keeping the same phase rotation in the three motors. Figure 24 is a phasor diagram of the electromotive forces and currents for the three motors. Electromotive forces of the three phases e_u, e_v and e_w consist of three components and are given by

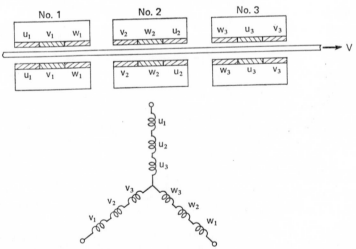

Fig. 23. Connection of three linear induction motors to eliminate unbalance due to end effect.

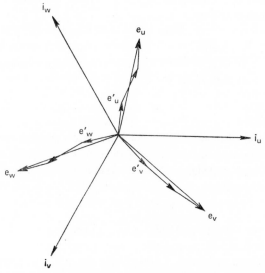

Fig. 24. Phasor diagram of emfs and currents of three linear induction motors shown in Fig. 23.

$$e_u = e_u' + e_v' \exp\left(j\frac{2}{3}\pi\right) + e_w' \exp\left(j\frac{4}{3}\pi\right)$$
$$= \sqrt{2}\, E_1,$$

$$e_v = e_v' + e_w' \exp\left(j\frac{2}{3}\pi\right) + e_u' \exp\left(-j\frac{4}{3}\pi\right)$$
$$= \sqrt{2}\, E_1 \exp\left(-j\frac{2}{3}\pi\right),$$

$$e_w = e_w' + e_u' \exp\left(j\frac{2}{3}\pi\right) + e_v \exp\left(j\frac{4}{3}\pi\right)$$
$$= \sqrt{2}\, E_1 \exp\left(-j\frac{4}{3}\pi\right). \tag{52}$$

They are now balanced as shown in Fig. 24. The series connection of three linear induction motors has significance for practical applications where many primary cores are installed in succession, such as for use in train propulsion. Our analysis of the linear induction motor under constant voltage is also based on the three balanced linear induction motors in Fig. 23. These motors together will hereafter be referred to as a balanced linear induction motor. When a balanced linear induction motor is supplied with a balanced three-phase voltage, the three-phase current is also balanced. The electromotive forces are given in Eq. (52), and their representative electromotive force per phase is denoted by E_1. The representative current of the balanced three-phase currents is denoted by I_1. For a given I_1 of i_u, i_v and i_w, we are now in a position to calculate the corresponding E_1. Then impedance for E_1 and I_1 is determined by

$$Z_{m2} = \frac{E_1}{I_1} \ . \tag{53}$$

Z_{m2} contains both the magnetizing impedance Z_m and the secondary impedance Z_2, connected in parallel, as shown in Fig. 25. Z_{2m} connected in series with primary impedance Z_1 gives terminal impedance Z_t, as shown in Fig. 25. Thus we have

$$Z_t = Z_1 + Z_{m2} \ . \tag{54}$$

After finding the terminal impedance Z_t of the linear induction motor which takes into account the end effect, motor performance under constant voltage is calculated as follows. For a given voltage V_1, the primary current I_1 is given by

$$I_1 = \frac{V_1}{Z_t} \ . \tag{55}$$

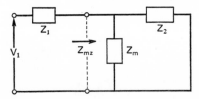

Fig. 25.   Equivalent circuit of a linear induction motor.

After finding the primary current I_1, the calculation of motor performance is the same as for performance under constant current drive which is explained in Chapter 5.

The characteristics of the three linear induction motors in the table in Appendix VI under constant voltage drive were calculated and the results are shown in Figs. 26, 27 and 28.

The general performance trend of a motor under constant voltage drive is similar to performance under constant current drive as described in Chapter 5. There is quite a distinction between low-speed motors (motor A) and high-speed motors (motors B and C). As Fig. 26(a) shows, the low-speed motor produces thrust at synchronous and higher speeds under constant voltage drive also; as in the case of constant current drive it is due to the end effect. As Fig. 26(c), (d) and (e) shows, in the small-slip region power factor and efficiency are improved due to the end effect, although improvement is slight.

In the high-speed motors, the influence of the end effect is opposite and adverse. As Figs. 27 and 28 show, in the low-slip region motor performance is considerably degraded; thrust, power factor, and efficiency being remarkably reduced.

In low-slip region the end effect weakens the normal magnetic field and keeps the primary current at higher value, resulting in a poor power factor and reduced thrust and efficiency. The working mechanism of the end effect will be explained in Chapter 7. The degradation of high-speed linear induction motor performance, exemplified in Figs. 27 and 28, is so extensive that the feasibility of a linear induction motor for high-speed application may be doubtful, if the end effect is left unremedied.

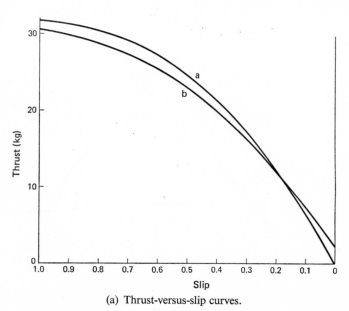

(a) Thrust-versus-slip curves.

Fig. 26. Characteristic curves for motor A (synchronous speed 9 m/s, gap length 15 mm, secondary 5 mm Al, 150 V). Curve *a*: without end effect; curve *b*: with end effect.

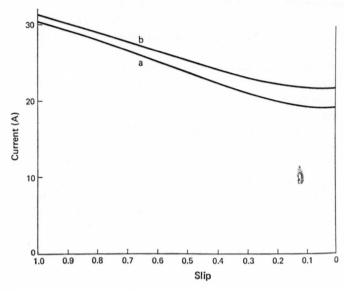

(b) Primary current-versus-slip curves.

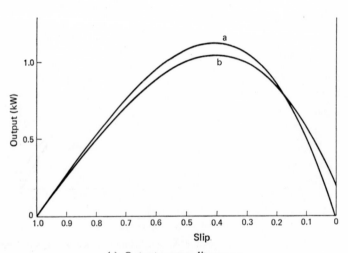

(c) Output-versus-slip curves.

Fig. 26.   Continued.

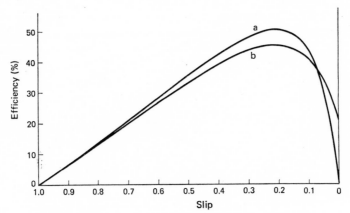

(d) Efficiency-versus-slip curves under constant voltage drive.

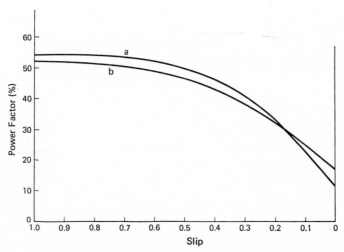

(e) Power factor-versus-slip curves under constant voltage drive.

Fig. 26. Continued.

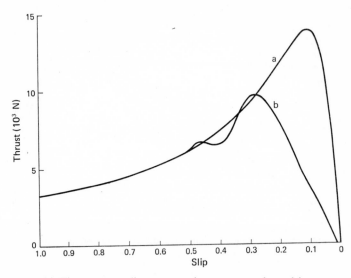

(a) Thrust-versus-slip curves under constant voltage drive.

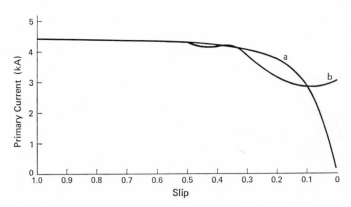

(b) Primary current-versus-slip curves under constant voltage drive.

Fig. 27.   Calculated performance curves for motor B (50 Hz, synchronous speed 150 m/s, 4 poles, 40 mm gap, secondary 10 mm Al). Curve *a*: without end effect; curve *b*: with end effect.

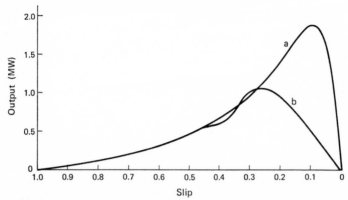

(c) Output-versus-slip curves under constant voltage drive.

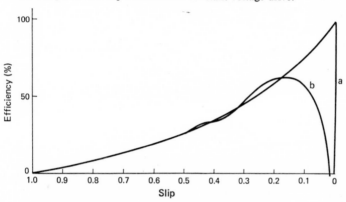

(d) Efficiency-versus-slip curves under constant voltage drive.

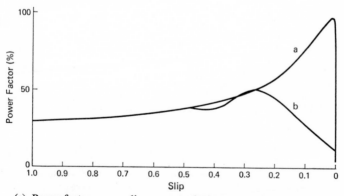

(e) Power factor-versus-slip curves under constant voltage drive.

Fig. 27. Continued.

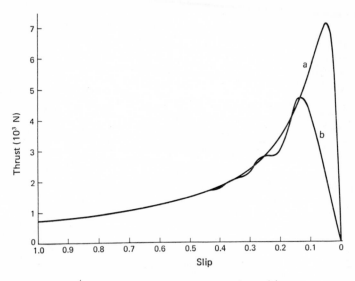

(a) Thrust-versus-slip curves under constant voltage drive.

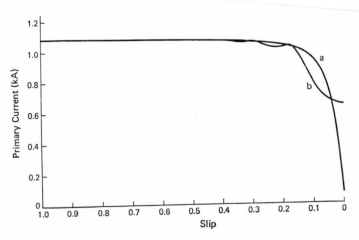

(b) Primary current-versus-slip curves under constant voltage drive.

Fig. 28. Calculated performance curves for motor C (150 Hz, synchronous speed 150 m/s, 10 poles, 40 mm gap, secondary 10 mm Al). Curve *a*: without end effect; curve *b*: with end effect.

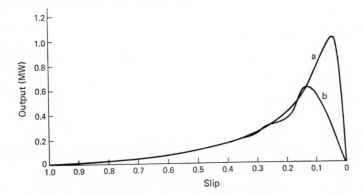

(c) Output-versus-slip curves under constant voltage drive.

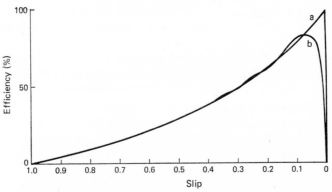

(d) Efficiency-versus-slip curves under constant voltage drive.

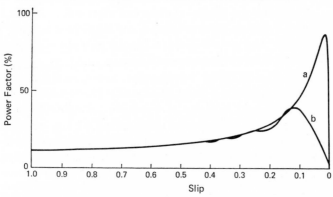

(e) Power factor-versus-slip curves under constant voltage drive.

Fig. 28. Continued.

Chapter 7

Low-speed and High-speed Motors*

It has been pointed out in several places in this book that low-speed and high-speed linear induction motors differ from each other in performance and characteristics. According to the value $\mu_0 v^2/4\omega\rho_s g$ given in Eq. (37) and the formula in Eq. (45), motors are divided into two groups. The dividing point is not so distinct and an intermediate zone exists. However, if $\mu_0 v^2/4\omega\rho_s g < 1$, the behavior of the motor is that of low-speed motor; and if $\mu_0 v^2/4\omega\rho_s g \gg 1$, the behavior is that of a high-speed motor.

In high-speed motors, length of penetration α_1 of the entry-end-effect wave is much longer than in low-speed motors as shown in Figs. 11 and 12, hence, the influence of the end-effect wave on motor performance is much greater in high-speed motors than in low-speed motors; this is a quantitative difference. As can be seen in Figs. 11 and 12, gap length g and surface resistivity ρ_s affect length of penetration α_1 of the entry-end-effect wave in low-speed and in high-speed regions differently. In the low-speed region the longer g is, the shorter α_1 becomes, while in the high-speed region the longer g is, the longer α_1 becomes. Influence of ρ_s is similar. This is the qualitative difference between high-speed and low-speed motors.

Curves of half-wave length τ_e of the end-effect wave in Figs. 15 and 16 have different tendencies in low-speed and high-speed regions. In the high-speed region, half-wave length τ_e is linearly proportional to the speed of the secondary and is independent from gap length g and secondary surface resistivity ρ_s. In the low-speed region, half-wave length τ_e is larger than the straight extension from the linearly proportional high-speed region and is dependent on g and ρ_s. These are qualitative differences.

Speed v_e of the end-effect wave is given in Eq. (36), and is proportional to τ_e. v_e-versus-v curves are given also in Figs. 15 and 16. In the high-speed region, speed v_e of the end-effect wave is the same as motor speed v, irrespective of supply frequency and also of gap length g and surface resis-

* Reported originally in refs. 5, 8, 12 and 13.

tivity ρ_S. In the low-speed region, v_e is higher than v and is dependent on f, g and ρ_S. For example, for 50 Hz and $\rho_S = 3.38 \times 10^{-5}$ Ω, v_e is 23 m/s at $v = 10$ m/s. The end-effect wave travels more than twice as fast as the motor speed in the low-speed region. When the slip is zero and the speed of the secondary conductive sheet is synchronous speed, the end-effect wave travels much faster than the synchronous speed. The super-synchronous speed of the end-effect wave at motor speed lower than synchronous speed occurs only in low-speed motors.

In low-speed motors the entry-end-effect wave attenuates rapidly and its influence on motor performance is comparatively small. In high-speed motors the entry-end-effect wave attenuates very slowly and its influence on motor performance is large. Moreover, the ways in which the end effect affects the performance of low-speed motors are quite different from the ways in which they affect the performance of high-speed motors. Figure 29 shows three ways in which the end effect affects the thrust of the linear induction motor. Figure 29(a) shows typical thrust-slip curves for low-speed motors. The end effect reduces thrust in the high-slip region, while it increases thrust in low-slip region, and positive thrust is produced even at synchronous and higher speeds; Fig. 19 is one example of such a case.

Figure 29(c) shows the typical thrust-slip characteristics for high-speed motors. In the high-slip region, the curve without the end effect and the curve with the end effect coincide with each other, while in the low-slip region, the end effect reduces thrust considerably. Thrust drops to zero at some positive slip and is negative at synchronous speed. The curve with the end effect undulates in middle slip range; Figs. 20 and 21 are examples of this case.

Figure 29(b) shows the typical thrust-slip characteristics of medium-speed motors. The end effect reduces thrust over entire region of slip 1–0.

Other characteristic curves, such as power factor-versus-slip curves and efficiency-versus-slip curves, are similar to the thrust-versus-slip curves in Fig. 29. It is very interesting in view of the theory of linear induction motors and very important in view of the application of linear induction motors to recognize the different ways in which the end effect affects linear induction motor performances. It seems that the recognition of this fact has been missing in most studies on linear induction motors.

Figures 27 and 28 show the extent to which thrust, power factor and efficiency are reduced by the end effect in high-speed motors. The degradation of motor performance is intolerably large and feasibility of a high-speed linear induction motor becomes doubtful.

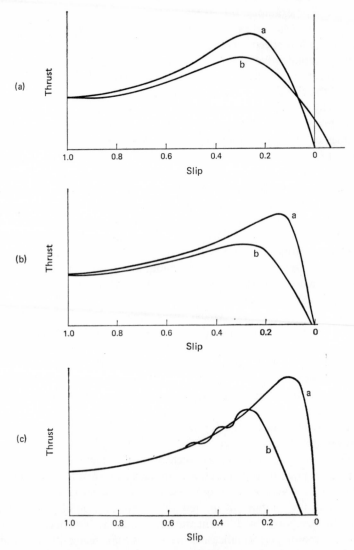

Fig. 29. Three kinds of thrust-slip characteristic curves. Curve *a*: without end effect; curve *b*: with end effect.

We shall now consider the reason why high-speed linear induction motors behave quite differently from low-speed linear induction motors.

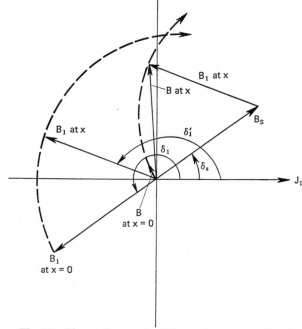

Fig. 30. Phasor diagram for high-speed motors at slip ≠ 0.

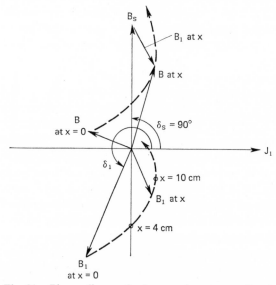

Fig. 31. Phasor diagram for low-speed motors at slip=0.

In the high-speed region, speed v_e of the end-effect wave, which is given in Eq. (40), is approximately equal to speed v of the secondary conductive sheet. Hence, when motor speed v is lower than its synchronous speed, v_e is always lower than the synchronous speed. As explained in Chapter 5, the influence of the exit-end effect wave B_2 in Eq. (35) is very small and can be neglected in most practical applications. Figures 30 and 31 show phasor diagrams of the normal wave B_S and the entry-end effect wave B_1 in Eq. (35). They shows phasors of the B_S wave and the B_1 wave and that of the primary current J_1. They are travelling waves which travel in the positive direction of the x coordinate. Both the B_S wave and the J_1 wave travel at synchronous speed and hence their relative phase relation remains the same as in Figs. 30 and 31 over the entire length of the iron core. Hence, phasors B_S and J_1 in Figs. 30 and 31 can represent the B_S wave and the J_1 wave for the entire length of the iron core.

On the other hand, the B_1 wave travels at a speed different from the synchronous speed and attenuates while travelling. The phasor of the B_1 wave moves in relation to those of the B_S wave and the J_1 wave, which are assumed to be stationary phasors. When the speed of the B_1 wave is lower than the synchronous speed, phasor B_1 moves clockwise, as in Fig. 30, and when speed of the B_1 wave is higher than the synchronous speed, phasor B_1 moves anti-clockwise, as in Fig. 31, as the B_1 wave travels in the positive direction of the x coordinate. The speed of the B_1 wave is in most cases lower than the synchronous speed. In high-speed motors, the speed v_e of the B_1 wave is given in Eq. (40) and is approximately equal to the motor speed v. Accordingly, in motor operation where the slip is between 1 and 0, the speed of the B_1 wave is always lower than synchronous speed and phasor B_1 rotates clockwise as shown in Fig. 30 as B_1 wave travels in the air gap. As explained above, in low-speed motors, speed v_e of the B_1 wave is higher than the motor speed v. Accordingly, it can happen that speed v_e of the B_1 wave becomes higher than synchronous speed when the slip is small. Then phasor B_1 rotates anti-clockwise as shown in Fig. 31, as the B_1 wave travels in a positive direction to the x coordinate.

The thrust generation between the J_1 wave and B waves will be explained now with respect to phasor diagrams in Figs. 30 and 31; total thrust is given by Eq. (47). Thrust per unit area generated between J_1 wave and B wave is given by $J_1 B^*/2$, where J_1 and B are phasors representing amplitudes and phases of the J_1 wave and the B wave, respectively, and B^* is a conjugate of phasor B. It is assumed that primary current j_1 is given and its phasor J_1 is fixed along the horizontal axis as shown in the phasor diagrams in Figs. 30 and 31. Then thrust per unit area between the

J_1 wave and the B wave is positive when phasor B lies in the right half plane, where phasor J_1 also lies, while the thrust is negative when phasor B lies in the left half plane. In Fig. 30, the thrust between the J_1 wave and the B_S wave is always positive since phasor B_S lies always in the right half plane. When it is integrated over the entire surface of the iron core, it provides total thrust when no end effect exists. Thrust between the J_1 wave and the B_1 wave is negative at the entry end since phasor B_1 lies in the left half plane for $x=0$. As the B_1 wave travels in the air gap, phasor B_1 rotates as shown in the phasor diagram. When phasor B_1 comes into the right half plane, thrust becomes positive. When the slip is greater, speed v_e of the B_1 wave is lower than synchronous speed, and then phasor B_1 rotates clockwise. Thrust generated between the J_1 wave and the B_1 wave alternates as the B_1 wave travels in the air gap. Thus, total thrust between the J_1 wave and the B_1 wave alternates as slip changes. When this thrust is superposed over the thrust between the J_1 wave and the B_S wave, the thrust-versus-slip curve undulates in the larger slip region, as Figs. 20 and 21 show. When the slip is small, speed v_e of the B_1 wave comes closer to synchronous speed, as Eq. (40) shows. Then phasor B_1 rotates clockwise slowly but it can not reach the right half plane. As a result thrust between the J_1 wave is negative. When the slip is very small, the speed of the B_1 wave is about the same as synchronous speed. Then phasor B_1 practically does not move and it weakens the B_S wave over the entire length of the iron core. This is the reason why thrust is lost, the power factor is remarkably reduced, efficiency is considerably reduced and the primary current remains large in the small-slip region, as characteristic curves of the high-speed linear induction motors, as shown in Figs. 27 and 28, indicate.

In low-speed motors, the situation in the high-slip region is similar to that of high-speed motors in which phasor B_1 rotates clockwise in the phasor diagram in Fig. 30. However, attenuation of the B_1 wave is so large that the B_1 wave becomes very small when it reaches the right half plane. Hence, the B_1 wave produces more negative thrust as it moves in the left half plane than the positive thrust it produces after it reaches the right half plane. Thus, the resultant thrust produced by the B_1 wave is negative in the high-slip region and it reduces the thrust curve with end effect from the corresponding curve without end effect in the high-slip region, as shown in Fig. 29(a). In low-slip region of low-speed motors, the situation is quite different. As was explained earlier in this chapter, speed v_e of the B_1 wave can be higher than synchronous speed and phasor B_1 rotates anti-clockwise as shown in Fig. 31. The starting point of phasor B_1 is located in the third quadrant close to the fourth quadrant. The B_1 wave then reaches the fourth quadrant very soon to produce positive thrust. This is the reason why thrust is produced even at synchronous speed and related performance is improved in the low-slip region of the low-speed linear induction motor, as Figs. 19 and 29(a) indicate.

It should be emphasized here that the differences between the performance of low-speed and high-speed motors are not only quantitative but qualitative. They are dependent on whether the speed of the B_1 wave is higher or lower than synchronous speed. They are fundamental differences.

Now that we understand the reason why performance in low-slip region of high-speed motors is extremely degraded due to the end effect, we may now be in a better position to eliminate the end effect. The way in which this can be done will be shown in Chapter 13.

Before leaving the discussion on differences between high-speed linear induction motors and low-speed linear induction motors, it is necessary and very useful to point out another difference between the two groups of linear induction motors.

It is generally believed that performance of the linear induction motor becomes poor because of the longer air gap. The air gap of the linear induction motor may be 40 to 50 times larger than that of the rotating induction motor. A longer air gap increases the exciting current, resulting in a poorer power factor and poorer efficiency. This may be true for low-speed motors, however, it is not necessarily true for high-speed motors. If there is no end effect, the power factor and efficiency of the high-speed linear induction motor can be very high, as shown by curve *a* in Figs. 27 and 28. It is the end effect that causes the poor power factor and efficiency of the high-speed linear induction motor. This situation has not been well understood, and many attempts have been made to improve the power factor of the linear induction motor without any attempt to eliminate the end effect. In most of these attempts, the use of composite secondary sheets, made of ferrous and nonferrous metals, were recommended.[1,20] Ferrous metals might provide a low-reluctance path to magnetic flux and nonferrous metals might provide a low-resistance path to a secondary current. However, the composite secondary is rather complicated in structure and is not so effective in improving performance, as will be explained in Section 10.3. Besides, if the end-effect is removed the high-speed linear induction motor achieve a very high power factor even with a simple nonferrous secondary sheet. This will be now explain.

The phase difference δ_s between the primary current J_1 and the normal flux wave B_s, as shown in phasor diagram of Fig. 30 are given in Eq. (16).

δ_S is also the phase difference between the primary induced voltage and the primary current. Cosine δ_S is the secondary power factor and is equal to the input power factor, if the primary impedance drop is neglected. Eq. (16) can be rewritten as

$$\delta_S = \tan^{-1} \frac{\pi \rho s g}{\mu_0 s \tau v_S} , \tag{56}$$

where s is slip and v_S is synchronous speed. In high-speed motors τv_S is very large and hence the gap length g can be longer, without increasing δ_S. If v_S is increased ten times, then g can be increased ten times for the same power factor. In the linear induction motor surface resistivity ρ_S is much smaller than the equivalent surface resistivity of the slotted rotor of the rotating induction motor. Surface resistivity of the secondary sheet may be 1/3 to 1/5 the time of the equivalent surface resistivity of the slotted rotor. Compared with the standard rotating induction motor, the numerical value $\rho_S / \tau v_S$ of the high-speed linear induction motor might be only 1/50 to 1/100. This means that without sacrificing the power factor, the gap length of the high-speed linear induction motor can be 50–100 times that of the low-speed rotating induction motor. In Figs. 27 and 28, the power factors of motors B and C are higher than 90% in spite of their longer gaps of 40 mm if no end effect exists. This fact is very important and confirms the essential necessity of suppressing the end effect of the high-speed linear induction motor.

Chapter 8

Measures to Alleviate the End Effect in High-speed Linear Induction Motor

8.1. SELECTION OF MOTOR PARAMETERS

Our theory and experiments revealed that the end effect exercises considerable influence on the linear motor performance, and that especially performance of the high-speed linear induction motor is degraded considerably. Without taking into account the influence of the end effect, performance and characteristics of the high-speed linear induction motor cannot be calculated with any accuracy of practical value. Our theory made it possible to calculate performance of the linear induction motor under the influence of the end effect, and it clarified detailed structure of the end effect. There are many factors which exercise influence on performance and characteristics of the linear induction motor, such as number of poles, gap length, frequency, secondary resistivity, impedances, and pole pitch, etc., and most of the factor also influence the end effect. The one-dimensional theory gave solutions for the air-gap field equation, which contain these parametric factors explicitly, and it is easier from the solutions of the one dimensional theory to determine parametric influence of each factor on the end effect. Although elimination or suppression of the end effect is desirable, alleviation of the adverse influence of the end effect is sufficient in many practical cases. Our one-dimensional theory revealed the general tendencies of the parametric influences on performance and characteristics of the high-speed linear induction motor under the influence of the end effect, and it provides guidance in proper selection of the parameters, which will alleviate the adverse influence of the end effect. The general tendencies of the parametric influences on the high-speed linear induction motor are as follows:

a) A larger number of poles reduces the slip range where the adverse influence of the end effect is considerable.

b) Higher secondary resistance reduces the length of penetration of the entry-end effect wave and alleviates the adverse influence of the end effect, although it degenerates the normal motor performance which is not under the influence of the end effect (see Fig. 12).

c) A larger air gap reduces the length of penetration of the entry-end wave and alleviate the adverse influence of the end effect, although it degenerates the normal motor performance which is not under the influence of the end effect (see Fig. 11). It is noted that the air-gap length g and the secondary surface resistivity ρ_s appear in the characteristic equation in Eq. (22) as the product $\rho_s g$, and that each of them exercises the same influence.

d) Higher frequency of the power supply reduces the length of penetration of the entry-end-effect wave and alleviates the adverse influence of the end effect (see Fig. 36), although it increases impedance drops and degrades the normal performance.

All the means to alleviate the adverse influence of the end effect in the high-speed linear induction motor tend to degrade the normal motor performance which is not under the influence of the end effect. It is necessary then to make compromise in selecting a suitable combination of the motor parameters. Under a given speed and output selection of higher frequency of the power supply is accompanied by a larger number of poles, and both tend to alleviate the adverse influence of the end effect. In addition, if the higher secondary resistance is chosen, further alleviation is expected. However, in many cases higher secondary resistance requires a very thin secondary conductive sheet, which is mechanically weak and has too little thermal capacity to withstand the secondary resistance loss increased by the higher secondary resistance. In such a case a sandwiched secondary plate, which consists of aluminum or copper sheets on both surfaces and a layer of different kind of material as shown in Fig. 63, may help. The kind of material used in the middle layer must increase mechanical strength and thermal capacity of the secondary plate and may be steel or nonmetallic. Or it might be necessary to cool the secondary with water to absorb and dissipate increased thermal loss.

When it is not possible to find a proper compromise, then it is necessary to eliminate or suppress the end effect. Methods of eliminating or suppressing the end effect will be explained in Chapters 13 and 14.

8.2. Selection of the Number of Poles[14]

It was suggested by Laithwaite[1] that a large number of poles helps to alleviate performance degradation, however, it seems that no sound reason for this suggestion has been given so far. In order to discuss measures to counter performance degradation it is necessary to derive qualitative expressions for performance under the influence of the end effect which have been lacking until now.

We will first derive formulas for calculating thrust reduction due to the end effect. As explained previously the influence of the B_2 wave on motor performance is small and can be neglected in most practical applications. The air-gap flux density in Eq. (35) then becomes

$$b = b_s + b_t$$

$$= B_s \exp j\left(\omega t - \frac{\pi}{\tau}x + \delta_s\right)$$

$$+ B_1 \exp\left(-\frac{x}{\alpha_1}\right) \exp j\left(\omega t - \frac{\pi}{\tau_e}x + \delta_1\right), \qquad (57)$$

then normal thrust, which is generated between the J_1 wave in Eq. (13) and the B_s wave in Eq. (57), is given by

$$F_n = D \int_0^L j_1 b_s dx$$

$$= D \int_0^L \mathrm{Re}\left[J_1 \exp j\left(\omega t - \frac{\pi}{\tau}x\right)\right]$$

$$\times \mathrm{Re}\left[B_s \exp j\left(\omega t - \frac{\pi}{\tau}x + \delta_s\right)\right] dx$$

$$= DJ_1 B_s \int_0^L \cos\left(\omega t - \frac{\pi}{\tau}x\right) \cos\left(\omega t - \frac{\pi}{\tau}x + \delta_s\right) dx$$

$$= \frac{DJ_1 B_s}{2} \int_0^L \left\{ \cos \delta_s + \cos\left(2\omega t - \frac{\pi}{\tau}x + \delta_s\right)\right\} dx$$

$$= \frac{DLJ_1 B_s}{2} \cos \delta_s . \qquad (58)$$

Thrust, which is generated between the J_1 wave and the B_1 wave of Eq. (57) is given by

$$F_e = D \int_0^L j_1 b_1 dx = D \int_0^L \mathrm{Re}\left[J_1 \exp j\left(\omega t - \frac{\pi}{\tau}x\right)\right] \times$$

$$\times \mathrm{Re}\left[B_1 \exp\left(-\frac{x}{\alpha_1} \right) \exp j\left(\omega t - \frac{\pi}{\tau_e}x + \delta_1 \right) \right] dx$$

$$= DJ_1 B_1 \int_0^L \exp\left(-\frac{x}{\alpha_1} \right) \cos\left(\omega t - \frac{\pi}{\tau}x \right) \cos\left(\omega t - \frac{\pi}{\tau_e}x + \delta_1 \right) dx$$

$$= \frac{DJ_1 B_1}{2}\left(\frac{1}{\left(\frac{1}{\alpha_1} \right)^2 + \left(\frac{\pi}{\tau}\frac{s}{1-s} \right)^2}\left[\frac{1}{\alpha_1}\left\{ \cos\delta_1 - \exp\left(\frac{L}{\alpha_1} \right)\cos\left(\delta_1 - \frac{\pi}{\tau}\frac{s}{1-s}L \right) \right\} \right. \right.$$

$$\left. + \frac{\pi}{\tau}\frac{s}{1-s}\left\{ \sin\delta_1 - \exp\left(-\frac{L}{\alpha_1} \right)\sin\left(\delta_1 - \frac{\pi}{\tau}\frac{s}{1-s}L \right) \right\} \right]$$

$$+ \frac{1}{\left(\frac{1}{\alpha_1} \right)^2 + \left(\frac{\pi}{\tau}\frac{2-s}{1-s} \right)^2}\left[\frac{1}{\alpha_1}\left\{ \cos(2\omega t + \delta_1) - \exp\left(-\frac{L}{\alpha_1} \right) \right. \right.$$

$$\times \cos\left(2\omega t - \frac{\pi}{\tau}\frac{2-s}{1-s}L + \delta_1 \right)\bigg\}$$

$$\left. \left. + \frac{\pi}{\tau}\frac{s}{1-s}\left\{ \sin(2\omega t + \delta_1) - \exp\left(-\frac{L}{\alpha_1} \right)\sin\left(2\omega t + \delta_1 + \frac{\pi}{\tau}\frac{2-s}{1-s}L \right) \right\} \right] \right) \quad (59)$$

In this derivation τ_e was substituted by Eq. (39), which is valid for high-speed motors. Hence Eq. (59) is valid for high-speed motors. The thrust F_e due to the end-effect wave B_1 contains double-frequency components, whose time average is zero. Neglecting the double-frequency components, thrust F_e is given by

$$F_e = \frac{DJ_1 B_1}{2}\left[\left(\frac{1}{\alpha_1} \right)^2 + \left(\frac{\pi}{\tau}\frac{s}{1-s} \right)^2 \right]^{-1}$$

$$\times \left[\frac{1}{\alpha_1}\left\{ \cos\delta_1 - \exp\left(-\frac{L}{\alpha_1} \right)\cos\left(\delta_1 - \frac{\pi}{\tau}\frac{s}{1-s}L \right) \right\} \right.$$

$$\left. + \frac{\pi}{\tau}\frac{s}{1-s}\left\{ \sin\delta_1 - \exp\left(-\frac{L}{\alpha_1} \right)\sin\left(\delta_1 - \frac{\pi}{\tau}\frac{s}{1-s}L \right) \right\} \right]. \quad (60)$$

Total thrust F generated is given by the sum of F_n in Eq. (58) and F_e in Eq. (60), and thus we have

$$F = F_n + F_e. \quad (61)$$

When motor speed is very high, the length of penetration α_1 becomes very large, as Figs. 11 and 12 show, and $B_1 \doteqdot B_S$ according to Eq. (44). For $(\alpha_1\pi/\tau)\{s/(1-s)\} \gg 1$, Eq. (60) now becomes

$$F_e = \frac{DJ_1 B_S \tau(1-s)}{2\pi s}\left\{ \sin\delta_1 - \sin\left(\delta_1 - \frac{\pi}{\tau}\frac{s}{1-s}L \right) \right\}. \quad (62)$$

From Eqs. (58) and (62) we get

$$\frac{F_e}{F_n} = \frac{\tau(1-s)}{L\pi s \cos \delta_s} \left\{ \sin \delta_1 - \sin\left(\delta_1 - \frac{\pi}{\tau} \frac{s}{1-s} L\right) \right\} . \tag{63}$$

Equation (63) indicates that end-effect thrust F_e varies sinusoidally as slip s changes. This produces thrust-slip curve undulation, as Figs. 20, 21, 27(a) and 28(a) show. Equation (63) also indicates that when the number of poles $P = L/\tau$ is large, the end-effect thrust F_e becomes smaller compared with normal thrust F_n.

Phase angle δ_1 of the B_1 wave is approximately equal to $\delta_s + \pi$, as Fig. 30 shows.

$$\delta_1 = \delta_s + \pi . \tag{64}$$

Substituting Eq. (64) into Eq. (63), we get

$$\frac{F_e}{F_n} = -\frac{\tau(1-s)}{L\pi s \operatorname{con} \delta_s} \left\{ \sin \delta_s - \sin\left(\delta_s - \frac{\pi}{\tau} \frac{s}{1-s} L\right) \right\} . \tag{65}$$

In Fig. 21, for example, curve b crosses curve a at several points. At these crossing points the end-effect thrust is zero and Eq. (65) becomes zero. At these crossing points, the following condition is satisfied:

$$\frac{\pi}{\tau} \frac{s}{1-s} L = 2\delta_s + (2m+1)\pi \tag{66}$$

or

$$\frac{\pi}{\tau} \frac{s}{1-s} L = 2m\pi , \tag{67}$$

where m is zero or a positive integer. After crossing the last point, which is nearest to the synchronous speed ($s=0$), the thrust curve drops rather abruptly. Hence, the last crossing point must not be too far from the synchronous speed. Otherwise, the slip range, where thrust curve b is considerably lower than the normal thrust given by curve a, becomes too wide and motor performance is degraded so much that the motor may lose its feasibility. In Eqs. (66) and (67), $L/\pi = P$ is the number of poles. In Eq. (66), $m=0$ gives smallest number of poles, which is given by

$$P = \frac{1-s}{\pi s}(2\delta_s + \pi) . \tag{68}$$

Even for phase angle $\delta_s = 0$, P is 20 for $s = 0.05$. This number of poles might be too large for practical applications.

As explained above, the location of the last crossing point on the normal characteristic curves is very important. On the thrust-versus-slip curve, the last crossing point should be located near the slip for maximum thrust, as shown in Fig. 32(b). If the last crossing point is located far to the left of the slip for maximum thrust as shown in Fig. 32(a), then maximum thrust for curve b, which includes the influence of the end effect, is reduced considerably and the degradation of thrust-versus-slip performance is extensive. An example of this is shown in Fig. 27(a).

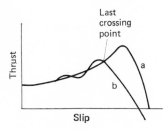

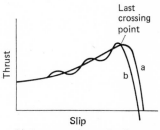

(a) Last crossing point lies left to the maximum thrust.

(b) Last crossing point lies near the maximum thrust.

Fig. 32. Thrust-versus-slip curves. Curve a: without end effect; curve b: with end effect.

Normal thrust due to the B_s wave is given in Eq. (58) and, substituting Eqs. (15) and (16), it is expressed as

$$F_n = \frac{J_1 B_{ST} \mu_0 \rho s (v_s - v)}{(\pi \rho g)^2 + \{\tau \mu_0 (v_s - v)\}^2} .$$ (69)

$dF_n/dv = 0$ gives

$$s_m = \frac{v_s - v}{v_s} = \frac{\tau \rho s g}{\tau \mu_0 v_s} .$$ (70)

This is the slip where maximum thrust occurs for constant current drive. In most practical high-speed linear induction motors slip s_m of Eq. (70) is smaller than 1%. The number of poles given by Eq. (68) for $s = s_m$ is too large to be of practical value. This indicates that some other means must be found to replace the method of selecting a larger number of poles. Figure 33 shows an example of the thrust per pole-versus-slip curves of linear induction motors whose number of poles is different and which are under constant current drive. The thrust per pole increases as the number of poles increases, and the trend is still absorbable even beyond 24 poles.

In cases of constant voltage drive slip s_m for maximum thrust is larger

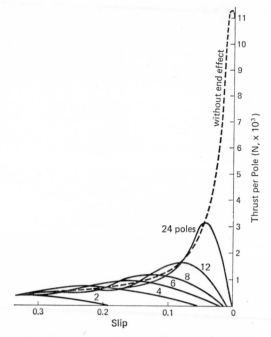

Fig. 33. Thrust per pole-versus-slip curves for motor B.

than s_m given by Eq. (70) for constant current drive. The method of selecting a large number of poles may be effective for alleviating the reduction
of thrust due to the end effect as explained above. However, similar discussions hold also for other performance parameters such as power factors, efficiency, etc. On the power factor-versus-slip curve and also the
efficiency-versus-slip curve, the last crossing point of the corresponding
curve with the end effect must be located near the slip for maximum value,
and must not be located far to the left of the slip. In most practical motors,
the slip for maximum power factor and efficiency is much smaller than
that for maximum thrust, and it is difficult to have enough large numbers
of poles in order to locate the last crossing point near the slip for maximum
value. This means that the degradation of the performance of most high-
speed linear induction motors due to the end effect is so extensive that if
some new means to eliminate the end effect is not found their feasibility
may be lost. Such means will be described in Chapters 13 and 14.

Chapter 9

Solutions for Field Equations of the Air Gap Based on a Two-dimensional Model*

9.1. INTRODUCTION

In the model of the linear induction motor in Fig. 8, spacial variations of physical quantities take place only in the x direction of the coordinate system. The magnetic field in the air gap is directed in the y direction; it has no component in the x direction and is a function of x and time t only. However, the actual magnetic field has also an x component and is a function of x, y and time t. Strictly speaking, it is also a function of z, the coordinate in the direction perpendicular to the x-y plane. Magnetic field density distribution is not uniform in the z direction and this nonuniformity of distribution has some influence on motor performance. This phenomenon will be analysed in Appendix I and the necessary correction, which takes into account its influence, will be given. When this correction is combined with the two-dimensional analysis in this chapter, the analysis becomes equivalent to a three-dimensional analysis.

The merits and demerits of the one-dimensional analysis and the two-dimensional analysis should be mentioned also. It is, of course, anticipated that the two-dimensional analysis gives better accuracy than the one-dimensional analysis in numerical evaluations of the physical quantities involved. However, as explained previously and as will be shown later, the one-dimensional analysis is accurate enough for many practical applications of linear induction motors. Since its analytical expressions are simpler and retain their parameters explicitly, it helps to obtain better insight into phenomena taking place within the linear induction motor. The one-dimensional analysis and the two-dimensional analysis have their own merits and demerits and supplement each other.

* Reported originally in refs. 12, 14, 16 and 27.

9.2. Derivation of Field Equations

Maxwell's equations are written for an electromagnetic field where displacement current can be neglected, as follows:

$$\nabla \times \mathbf{H} = \mathbf{J} , \tag{71}$$

$$\nabla \times \mathbf{E} = -\frac{\partial \mathbf{B}}{\partial t} , \tag{72}$$

$$\nabla \cdot \mathbf{B} = 0, \qquad \nabla \cdot \mathbf{D} = 0 . \tag{73}$$

For an isotropic medium moving with velocity $\mathbf{V}$, which is much smaller than velocity of light, the following equations hold:

$$\mathbf{J} = \sigma(\mathbf{E} + \mathbf{V} \times \mathbf{B}) , \tag{74}$$

$$\mathbf{B} = \mu \mathbf{H} . \tag{75}$$

In these equations, the symbols stand for the following vectors: $\mathbf{B}$, magnetic flux density, $\mathbf{H}$, magnetic field intensity, $\mathbf{D}$, dielectric flux density, $\mathbf{E}$, electric field intensity, and $\mathbf{J}$, electric current density. μ indicates permeability and σ conductivity.

Introducing the vector potential $\boldsymbol{\Phi}$, the following equations are added:

$$\mathbf{B} = \nabla \times \boldsymbol{\Phi} , \tag{76}$$

$$\mathbf{E} = -\frac{\partial \boldsymbol{\Phi}}{\partial t} . \tag{77}$$

If we choose vector potential $\boldsymbol{\Phi}$ for which the following equation holds:

$$\nabla \cdot \boldsymbol{\Phi} = 0 , \tag{78}$$

then from Eqs. (71) and (76) we get

$$\nabla^2 \boldsymbol{\Phi} = -\mu \mathbf{J} .$$

Substituting $\mathbf{J}$ from Eq. (74) gives

$$\nabla^2 \boldsymbol{\Phi} = \mu\sigma \left\{ \frac{\partial \boldsymbol{\Phi}}{\partial t} - \mathbf{V} \times (\nabla \times \boldsymbol{\Phi}) \right\} . \tag{79}$$

This is the general field equation in terms of vector potential $\boldsymbol{\Phi}$.

The model for the two-dimensional analysis of the linear induction motor is shown in Fig. 34. The coordinate axes are chosen as indicated in the figure and the three different regions are denoted by different indices. The iron core is region 1, the secondary conductive sheet is region 2, and

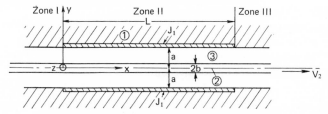

Fig. 34. Model for two-dimensional analysis of a linear induction motor.

the air gap is region 3. Vector potential Φ in region 1 is Φ_1, that in region 2 is Φ_2 and that in region 3 is Φ_3, etc. We shall now derive field equations for each of the three regions from the general field equation of Eq. (79).

In the model shown in Fig. 34, the primary iron cores are extended to infinity in both directions of the x coordinate, and the upper core is extended to plus infinity of the y coordinate and the lower core is extended to minus infinity of the y coordinate. But the primary current sheets are limited and exit between point $x=0$ and point $x=L$. Discontinuities of the primary current sheet are taken into consideration, while those of the iron cores are not. End effect is, therefore, caused only by the primary current but not by the iron cores. In actual linear induction motors, dimensions of the iron cores are also limited. End effect caused only by the end of the primary current and end effect caused by ends of both the primary current and iron core should be different from each other.

But actually the unlimited extension of the primary iron cores in the assumption causes appreciable error neither in the numerical evaluation of the field distribution in the air gap, as will be shown in Section 10.2, nor in the calculation of the linear induction motor performance, as will be shown in Chapter 11. The theoretical base for the validity of the assumption lies in the fact that the backward travelling wave attenuate so quickly, as was shown in Chapter 4, that practically no magnetic field is excited by the primary current in the air-gap zone in front of the entry end (zone I in Fig. 34). This means that the boundary condition with regard to the magnetic field in the vicinity of the entry end of the air gap is practically unchanged by the assumption. It should be pointed out also that the boundary condition in the vicinity of the exit end is changed appreciably by the assumption. This gives rise to error in the calculation of the backward travelling wave caused by the exit end; however, as mentioned above, it attenuates so quick that it has no significant influence on the motor performance. Accordingly, it can be concluded that the assumption is a sound one and this fact will be confirmed in all the calculations and experiments that follow. It should be pointed out here that the case where both primary irons and windings are limited is treated in Appendix II and the case where there is no fringing at both ends is treated in Appendix IV.

In the model shown in Fig. 34, the primary current $\mathbf{J}_1$ flows in the z direction and hence $\mathbf{B}$ and $\mathbf{H}$ have only an x component and a y component but no z component. Thus the model reveals the two-dimensional problem in the x-y plane. In this two-dimensional problem, vector potential $\boldsymbol{\Phi}$ has a z component only and will be denoted as ϕ instead of ϕ_z.

Then Eq. (76) gives the following equations:

$$B_x = \frac{\partial \phi}{\partial y}, \qquad B_y = -\frac{\partial \phi}{\partial x}. \tag{80}$$

In region 2 Eq. (79) becomes

$$\frac{\partial^2 \phi_2}{\partial x^2} + \frac{\partial^2 \phi_2}{\partial y^2} = \mu_2 \sigma_2 \left(\frac{\partial \phi_2}{\partial t} + v \frac{\partial \phi_2}{\partial x} \right). \tag{81}$$

Here v_2 is speed of the secondary sheet. In regions 1 and 3, where σ is zero, this equation is reduced to the following Laplace's equations:

$$\frac{\partial^2 \phi_1}{\partial x^2} + \frac{\partial^2 \phi_1}{\partial y^2} = 0, \tag{82}$$

$$\frac{\partial^2 \phi_3}{\partial x^2} + \frac{\partial^2 \phi_3}{\partial y^2} = 0. \tag{83}$$

9.3. Boundary Conditions

At the boundary between region 1 (iron core) and region 2 (air gap) there lies current sheet j_1, and the following relation holds:

$$\frac{B_{1x}}{\mu_1} - \frac{B_{2x}}{\mu_2} = j_1 \qquad \text{for} \quad 0 < x < L, \quad y = \pm a, \tag{84}$$

where index x denotes the x component. In terms of the vector potential given in Eq. (80), this equation is written as follows:

$$\frac{1}{\mu_1} \frac{\partial \phi_1}{\partial y} - \frac{1}{\mu_2} \frac{\partial \phi_2}{\partial y} = j_1 \qquad \text{for} \quad 0 < x < L, \quad y = \pm a. \tag{85}$$

Since j_1 is zero for $x < 0$ (zone I) and $x > L$ (zone III), Eq. (85) becomes

$$\frac{1}{\mu_1} \frac{\partial \phi_1}{\partial y} - \frac{1}{\mu_2} \frac{\partial \phi_2}{\partial y} = 0 \qquad \text{for} \quad x < 0 \quad \text{and} \quad x > L, \quad y = \pm a. \tag{86}$$

The continuity of B_y at the boundary gives

$$B_{1y} = B_{2y} \quad \text{for} \quad y = \pm a. \tag{87}$$

In terms of vector potential this equation becomes

$$\frac{\partial \phi_1}{\partial x} = \frac{\partial \phi_2}{\partial x} \quad \text{for} \quad y = \pm a. \tag{88}$$

At the boundary between region 2 and region 3 continuity of H_x gives

$$\frac{1}{\mu_2} \frac{\partial \phi_2}{\partial y} = \frac{1}{\mu_3} \frac{\partial \phi_3}{\partial y} \quad \text{for} \quad y = \pm b. \tag{89}$$

And continuity of B_y at the boundary gives

$$\frac{\partial \phi_2}{\partial x} = \frac{\partial \phi_3}{\partial x} \quad \text{for} \quad y = \pm b. \tag{90}$$

9.4. SOLUTION BY MEANS OF FOURIER TRANSFORMS

The vector potentials in the two-dimensional model have only z components and functions of x, y and time, and can be denoted by $\phi(x,y,t)$. With respect to time, $\phi(x,y,t)$ is stationary alternating, and the following form of separated variables can be properly adopted.

$$\phi(x,y,t) = \phi(x,y)\varepsilon^{j\omega t}. \tag{91}$$

Substitution of Eq. (91) into ϕ_2 in Eq. (81) gives

$$\left(\nabla^2 - j\omega\mu_2\sigma_2 - v\mu_2\sigma_2 \frac{\partial}{\partial x} \right) \phi_2 = 0 . \tag{92}$$

Let the Fourier transform of $\phi(x,y)$ with respect to x be denoted by $\tilde{\phi}(\xi,y)$, which is then given by

$$\tilde{\phi}(\xi,y) = \int_{-\infty}^{\infty} \phi(x,y)\varepsilon^{-j\xi x} dx . \tag{93}$$

Equation (92) in terms of a Fourier transform is given by

$$\left(-\xi^2 + \frac{d^2}{dy^2} - j\omega\mu_2\sigma_2 - jv\,\mu_2\sigma_2\xi \right) \tilde{\phi}_2(\xi,y) = 0 . \tag{94}$$

Hence we have

$$\frac{d^2\tilde{\phi}_2}{dy^2} = (\xi^2 + jv\mu_2\sigma_2\xi + j\omega\mu_2\sigma_2)\tilde{\phi} \equiv \gamma^2\tilde{\phi}_2 , \tag{95}$$

where the following abbreviation was introduced:

$$\xi^2 + jv\mu_2\sigma_2\xi + j\omega\mu_2\sigma_2 \equiv \gamma^2 . \tag{96}$$

Equations (82) and (83), when expressed in Fourier transforms, become

$$\frac{d^2\tilde{\phi}_1}{dy^2} = \xi^2\tilde{\phi}_1 , \tag{97}$$

$$\frac{d^2\tilde{\phi}_3}{dy^2} = \xi^2\tilde{\phi}_3 . \tag{98}$$

Now Eqs. (95), (97) and (98) will be solved.

Clearly, the field is symmetric with respect to the x coordinate axis, and hence, ϕ_2 and $\tilde{\phi}_2$ are even functions of y. In view of Eq. (95), it is assumed that $\tilde{\phi}_2$ takes the following form:

$$\tilde{\phi}_2 = C \cosh{(\gamma y)} . \tag{99}$$

$\tilde{\phi}_1$ in the iron core must tend to zero as y goes to infinity. Then Eq. (97) suggests the following form for $\tilde{\phi}_1$:

$$\tilde{\phi}_1 = D\varepsilon^{-\xi y}. \tag{100}$$

The solution of Eq. (98) may have the following general form:

$$\tilde{\phi}_3 = M\varepsilon^{\xi y} + N\varepsilon^{-\xi y}. \tag{101}$$

C, D, M and N in these equations can be determined from the boundary conditions. The boundary conditions given in Eqs. (85) and (86) are Fourier transformed by the next equation.

$$\int_{-\infty}^{\infty} \left(\frac{1}{\mu_1} \frac{\partial\phi_1}{\partial y} - \frac{1}{\mu_2} \frac{\partial\phi_2}{\partial y} \right) \varepsilon^{-j\xi x} dx = \int_{-\infty}^{\infty} j_1 \varepsilon^{-j\xi x} dx . \tag{102}$$

This is rewritten as

$$\frac{1}{\mu_1} \frac{\partial\tilde{\phi}_1}{\partial y} - \frac{1}{\mu_2} \frac{\partial\tilde{\phi}_2}{\partial y} = \tilde{J}_1 \qquad \text{for} \quad y = \pm a . \tag{103}$$

And $\tilde{J}_1$ is the Fourier transform of primary current j_1 and is given by

$$\tilde{J}_1 = \int_{-\infty}^{\infty} J_1\varepsilon^{-jkx}\varepsilon^{-j\xi x} dx$$

$$= \int_{0}^{L} J_1\varepsilon^{-jkx}\varepsilon^{-j\xi x} dx$$

$$= \frac{jJ_1}{\xi+k} \{\varepsilon^{-j(\xi+k)L} - 1\} , \tag{104}$$

where

$$k = \pi/\tau \ . \tag{105}$$

Similarly, the boundary condition of Eq. (88) becomes a Fourier transform in Eq. (106).

$$\frac{1}{\mu_2} \frac{\partial \tilde{\phi}_2}{\partial y} - \frac{1}{\mu_3} \frac{\partial \tilde{\phi}_3}{\partial y} = 0 \qquad \text{for} \quad y = \pm b \ . \tag{106}$$

The boundary conditions of Eqs. (88) and (90) are transformed into Fourier transforms giving, respectively,

$$\tilde{\phi}_1 = \tilde{\phi}_2 \qquad \text{for} \quad y = \pm a \ , \tag{107}$$

$$\tilde{\phi}_2 = \tilde{\phi}_3 \qquad \text{for} \quad y = \pm b \ . \tag{108}$$

From the boundary condition of Eqs. (103), (106), (107) and (108), C of Eq. (99), and M and N of Eq. (101) are determined. Here permeability μ of the iron core is assumed to be infinity. The results are given below

$$C = \frac{\mu_3 \tilde{J}_1}{\xi \cosh b\gamma \sinh(a-b)\xi + \frac{\mu_3}{\mu_2}\gamma \sinh b\gamma \cosh(a-b)\xi} \ , \tag{109}$$

$$M = \frac{C}{2}\left(\cosh b\gamma + \frac{\mu_3}{\mu_2}\frac{\gamma}{\xi}\sinh b\gamma\right)\varepsilon^{-\xi b}, \tag{110}$$

$$N = \frac{C}{2}\left(\cosh b\gamma - \frac{\mu_3}{\mu_2}\frac{\gamma}{\xi}\sinh b\gamma\right)\varepsilon^{\xi b}. \tag{111}$$

Substituting C, M and N into Eqs. (99) and (101), the Fourier transforms of vector potentials are obtained as follows:
In region 2 (the secondary conductive sheet)

$$\tilde{\phi}_2(x,y) = \frac{\mu_3 \tilde{J}_1 \cosh \gamma y}{\xi \cosh b\gamma \sinh(a-b)\xi + \frac{\mu_3}{\mu_2}\gamma \sinh b\gamma \cosh(a-b)\xi} \ . \tag{112}$$

In region 3 (the air gap)

$$\tilde{\phi}_3(x,y) = \frac{\mu_3 \tilde{J}_1 \left\{\cosh b\gamma \cosh(y-b)\xi + \frac{\mu_3}{\mu_2}\frac{\gamma}{\xi}\sinh b\gamma \sinh(y-b)\xi\right\}}{\xi \cosh b\gamma \sinh(a-b)\xi + \frac{\mu_3}{\mu_2}\gamma \sinh b\gamma \cosh(a-b)\xi} \ . \tag{113}$$

Inversely transforming the Fourier transforms of Eqs. (112) and (113), the vector potentials are obtained as follows:

In region 2 (the secondary sheet)

$$\phi_2(x,y)=\frac{1}{2\pi}\int_{-\infty}^{\infty}\frac{\mu_3\tilde{J}_1\cosh\gamma y}{\xi\cosh b\gamma\sinh(a-b)\xi+\frac{\mu_3}{\mu_2}\gamma\sinh b\gamma\cosh(a-b)\xi}\,\varepsilon^{j\xi x}d\xi \quad (114)$$

$$\text{for}\quad -b\le y\le b.$$

In region 3 (the air gap)

$$\phi_3(x,y)=\frac{1}{2\pi}\int_{-\infty}^{\infty}\frac{\mu_3\tilde{J}_1\left\{\cosh b\gamma\cosh(y-b)\xi+\frac{\mu_3}{\mu_2}\frac{\gamma}{\xi}\sinh b\gamma\sinh(y-b)\xi\right\}\varepsilon^{j\xi x}}{\xi\cosh b\gamma\sinh(a-b)\xi+\frac{\mu_3}{\mu_2}\gamma\sinh b\gamma\cosh(a-b)\xi}\,d\xi$$

$$\text{for}\quad b\le|y|\le a. \quad (115)$$

These are vector potentials in the air gap and in the secondary sheet, respectively, which provide the electromagnetic field in the *x-y* plane. In these equations, $\tilde{J}_1$ is the Fourier transform of the primary current and is given by Eq. (104). Inserting Eq. (104) into Eqs. (114) and (115), we have:
In region 2 (the secondary sheet)

$$\phi_2(x,y)=\frac{j\mu_3J_1}{2\pi}\int_{-\infty}^{\infty}\frac{\{\varepsilon^{-j(\xi+k)L}-1\}\cosh\gamma y}{(\xi+k)H(\xi)}\varepsilon^{j\xi x}d\xi . \quad (116)$$

In region 3 (the air gap)

$$\phi_3(x,y)=\frac{j\mu_3J_1}{2\pi}\int_{-\infty}^{\infty}\frac{\{\varepsilon^{-j(\xi+k)L}-1\}\,G(\xi,y)}{(\xi+k)H(\xi)}\varepsilon^{j\xi x}d\xi . \quad (117)$$

Here $G(\xi,y)$ and $H(\xi)$ are given by

$$G(\xi,y)=\cosh b\gamma\cosh(y-b)\xi+\frac{\mu_3}{\mu_2}\frac{\gamma}{\xi}\sinh b\gamma\sinh(y-b)\xi , \quad (118)$$

$$H(\xi)=\xi\cosh b\gamma\sinh(a-b)\xi+\frac{\mu_3}{\mu_2}\gamma\sinh b\gamma\cosh(a-b)\xi . \quad (119)$$

Integrations in Eqs. (116) and (117) will be now carried out by the residue theorem. One of the poles of the integrands of Eqs. (116) and (117) is given by

$$\xi=-k. \quad (120)$$

And the other poles are given as roots of the following equations:

$$H(\xi) = \xi \cosh b\gamma \sinh(a-b)\xi + \frac{\mu_3}{\mu_2}\gamma \sinh b\gamma \cosh(a-b)\xi = 0 , \quad (121)$$

$$\gamma^2 = \xi^2 + jv\mu_2\sigma_2\xi + j\omega\mu_2\sigma_2 . \quad (122)$$

Here Eq. (122) comes from Eq. (96). It is not possible to derive analytical expressions of roots of Eqs. (121) and (122), but it is possible to determine the roots by numerical calculation with a computer. As will be seen below, there are two important roots, which are located near the origin of the ξ plane (the complex plane), and other roots, if any, are far from the origin and terms controlled by these roots attenuate very quickly. Accordingly, two roots are sufficient for practical applications. Since they are not large in magnitude, in Eqs. (121) and (122) $|(a-b)\xi| \ll 1$ and $|b\gamma| \ll 1$ in most practical cases. Then Eq. (121) is approximated by

$$H(\xi) \doteq \xi(a-b)\xi + \frac{\mu_3}{\mu_2}b\gamma^2 = \left\{a + \left(\frac{\mu_3}{\mu_2}-1\right)b\right\}\xi^2 + jbv\mu_2\sigma_2\xi + jb\omega\mu_2\sigma_2$$
$$= 0 . \quad (123)$$

This equation is of second order with respect to ξ and let its two roots be denoted by ξ_0 and ξ_0'. Then Eq. (123) can be written as

$$H(\xi) \doteq \left\{a + \left(\frac{\mu_3}{\mu_2}-1\right)b\right\}(\xi-\xi_0)(\xi-\xi_0') . \quad (124)$$

ξ_0 and ξ_0' are approximate values of poles of the integrands of Eqs. (116) and (117). The exact values of the poles can be searched by the Newton-Raphson method with a computer, using ξ_0 and ξ_0' in Eq. (124) as starting values. In many practical cases, especially for high-speed motors, errors of ξ_0 and ξ_0' in Eq. (124) are not large.

Having found the poles of the integrands of Eqs. (116) and (117), we shall now apply the residue theorem. Vector potential $\phi_3(x,y)$ in Eq. (117) is modified as follows:

$$\phi_3(x,y) = \frac{j\mu_3 J_1}{2\pi} \int_{-\infty}^{\infty} \frac{\varepsilon^{j(x-L)\xi}\varepsilon^{-jkL} - \varepsilon^{j\xi x}}{\xi+k} \cdot \frac{G(\xi,y)}{H(\xi)} d\xi . \quad (125)$$

In the integrand $G(\xi,y)/(\xi+k)H(\xi)$ is analytic except at the three poles shown in Fig. 35 and tends to zero as $|\xi|$ goes to infinity. Hence, the residue theorem is applicable. Of the two closed paths of integration in the figure proper one must be chosen so that the integration along the infinite semicircle does not make any contribution. Then the exponential terms of the form $\varepsilon^{j\xi d}$ in the integrand must vanish as $|\xi|$ goes to infinity. Hence, for $d<0$ the lower semicircle must be the path of integration and for $d>0$

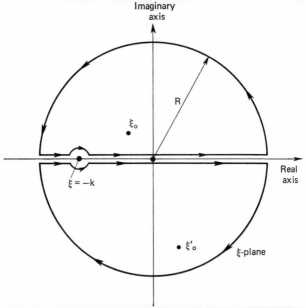

Fig. 35. Location of poles and paths of integration in ξ-plane.

the upper semicircle must be the path of integration. For integration of Eq. (125), then, the path of integration must be the lower semicircle if $x < 0$, and it must be the upper semicircle if $x > L$. If $0 < x < L$, the path must be the lower semicircle for the first term in the bracket and the upper semicircle for the second term in the bracket. Then the residue theorem gives the following result for the integration of Eq. (125).

In zone I $(x < 0)$

$$\phi_3(x,y) = \mu_3 J_1 \frac{\exp\{-j(\xi_0' + k)L\} - 1}{\xi_0' + k} \cdot \frac{G(\xi_0', y)}{H'(\xi_0')} \exp(j\xi_0' x) . \qquad (126)$$

In zone II $(0 < x < L)$

$$\phi_3(x,y) = \mu_3 J_1 \left[\frac{G(-k,y)}{H(-k)} \exp(-jkx) + \frac{G(\xi_0,y)}{(\xi_0 + k)H'(\xi_0)} \exp(j\xi_0 x) \right.$$

$$\left. + \frac{\exp\{-j(k + \xi_0')L\}}{(\xi_0' + k)} \cdot \frac{G(\xi_0,y)}{H'(\xi_0')} \exp(j\xi_0' x) \right] . \qquad (127)^*$$

---

* The first term comes from the integration along the infinitesimal semicircles around the pole $\xi = -k$ in Fig. 35.

In zone III ($L < x$)

$$\phi_3(x,y) = \mu_3 J_1 \frac{1 - \exp\{-j(k+\xi_0)L\}}{\xi_0 + k} \cdot \frac{G(\xi_0,y)}{H'(\xi_0)} \exp(j\xi_0 x) . \qquad (128)$$

Here $H'(\xi)$ is the first derivative of $H(\xi)$, and $G(\xi,y)$ is given by Eq. (118).

Vector potentials in the secondary sheet are also given by Eqs. (126)–(128), when $G(\xi,y)$ are replaced with cosh γy.

The y component of the magnetic flux density in the air gap is important because it is the component which produces thrust in combined action with the primary current j_1. It is given by $B_y = -\partial\phi_3/\partial x$ in Eq. (80), and from Eqs. (126)–(128) we get:

In zone I ($x < 0$)

$$B_{3y}(x,y) = -j\xi_0'\mu_0 J_1 \frac{\exp\{-j(\xi_0'+k)L\} - 1}{\xi_0' + k} \cdot \frac{G(\xi_0'y,)}{H'(\xi_0')} \exp(j\xi_0'x) . \qquad (129)$$

In zone II ($0 < x < L$)

$$B_{3y}(x,y) = jk\mu_3 J_1 \frac{G(-k,y)}{H(-k)} \exp(-jkx)$$

$$- j\xi_0\mu_3 J_1 \frac{G(\xi_0,y)}{(\xi_0+k)H'(\xi_0)} \exp(j\xi_0 x)$$

$$- j\xi_0'\mu_3 J_1 \frac{\exp\{-j(\xi_0'+k)L\}}{(\xi_0'+k)} \cdot \frac{G(\xi_0',y)}{H'(\xi_0')} \exp(j\xi_0'x) . \qquad (130)$$

In zone III ($L < x$)

$$B_{3y}(x,y) = j\xi_0\mu_3 J_1 \frac{\exp\{-j(\xi_0+k)L\} - 1}{\xi_0 + k} \cdot \frac{G(\xi_0,y)}{H'(\xi_0)} \exp(j\xi_0 x) . \qquad (131)$$

In the case where the secondary sheet is nonmagnetic, permeabilities are the same in the air gap and in the secondary sheet and we have $\mu_2 = \mu_3 = \mu_0$. Then Eqs. (118) and (119) become

$$G_0(\xi,y) = \cosh b\gamma \, \cosh(y+b)\xi + (\gamma/\xi) \sinh b\gamma \, \sinh(y-b)\xi , \qquad (132)$$

$$H_0(\xi) = \xi \cosh b\gamma \, \sinh(a-b)\xi + \gamma \sinh b\gamma \, \cosh(a-b)\xi . \qquad (133)$$

And Eq. (124) becomes

$$H_0(\xi) \doteq a(\xi - \xi_0) \, (\xi - \xi_0') . \qquad (134)$$

For the nonmagnetic secondary vector potentials in Eqs. (126)–(128) are expressed as follows:

In zone I $(x<0)$

$$\phi_3(x,y)=\mu_0 J_1 \frac{A(\xi_0')}{B(\xi_0')} \cdot \frac{\exp\{-j(k+\xi_0')L\}-1}{\xi_0'+k} \exp(j\xi_0'x). \tag{135}$$

In zone II $(0<x<L)$

$$\phi_3(x,y)=\mu_0 J_1 \frac{\cosh(y-b)k+\frac{\gamma_k}{k}\tanh b\gamma_k \sinh(y-b)k}{\{k\tanh(a-b)k+\gamma_k\tanh b\gamma_k\}\cosh(a-b)k}\exp(-jkx)$$

$$+\mu_0 J_1 \frac{A(\xi_0)}{B(\xi_0)} \cdot \frac{1}{\xi_0+k}\exp(j\xi_0 x)$$

$$+\mu_0 J_1 \frac{A(\xi_0')}{B(\xi_0')} \cdot \frac{\exp\{-j(k+L)\xi_0'\}}{\xi_0'+k}\exp(j\xi_0'x) , \tag{136}$$

In zone III $(L<x)$

$$\phi_3(x,y)=\mu_0 J_1 \frac{A(\xi_0)}{B(\xi_0)} \cdot \frac{1-\exp\{-j(k+\xi_0)L\}}{\xi_0+k}\exp(j\xi_0 x) . \tag{137}$$

Here

$$A(\xi)=\cosh(y-b)\xi+\frac{\gamma}{\xi}\tanh b\gamma \sinh(y-b)\xi ,$$

$$B(\xi)=\left\{\tanh(a-b)\xi+(a-b)\xi \operatorname{sech}^2(a-b)\xi+\left(\xi+\frac{jv\mu_0\sigma}{2}\right)\frac{\tanh b\gamma}{\gamma}\right.$$

$$\left.+b\left(\xi+\frac{jv\mu_0\sigma}{2}\right)\operatorname{sech}^2 b\gamma\right\}\cosh(a-b)\xi ,$$

and γ_0 is the value of γ with $\xi=\xi_0$, γ_0' is the value of γ with $\xi=\xi_0'$ and γ_k is the value γ with $\xi=-k$ in Eq. (122).

For the nonmagnetic secondary, the y components of magnetic flux density in the air gap are obtained from the relation $B_y=-\partial\phi_3/\partial x$ and Eqs. (135)–(137).

They are expressed as follows:
In zone I $(x<0)$

$$B_{3y}(x,y)=-j\xi_0'\mu_0 J_1 \frac{A(\xi_0')}{B(\xi_0')} \cdot \frac{\exp\{-j(k+\xi_0')L\}-1}{\xi_0'+k}\exp(j\xi_0'x) . \tag{138}$$

In zone II $(0 < x < L)$

$$B_{3y}(x,y) = jk\mu_0 J_1 \frac{\cosh(y-b)k + \frac{\gamma_k}{k}\tanh b\gamma_k \sinh(y-b)k}{\{k\tanh(a-b)k + \gamma_k \tanh b\gamma_k\}\cosh(a-b)k}\exp(-jkx)$$

$$- j\xi_0\mu_0 J_1\frac{A(\xi_0)}{B(\xi_0)} \cdot \frac{1}{\xi_0 + k}\exp(j\xi_0 x)$$

$$- j\xi_0'\mu_0 J_1\frac{A(\xi_0')}{B(\xi_0')} \cdot \frac{\exp\{-j(k+L)\xi_0'\}}{\xi_0' + k}\exp(j\xi_0 x) . \tag{139}$$

In zone III $(L < x)$

$$B_{3y}(x,y) = - j\xi_0\mu_0 J_1\frac{A(\xi_0)}{B(\xi_0)} \cdot \frac{1 - \exp\{-j(k+\xi_0)L\}}{\xi_0 + k}\exp(j\xi_0 x) . \tag{140}$$

Each term of the vector potentials or the y components of the magnetic flux density in Eqs. (126)–(128), (129)–(131) and (138)–(140) contains an exponential function of x, multiplied by one of the three poles, $-k$, ξ_0 and ξ_0'. As shown in Fig. 35, $-k$ is a negative real, ξ_0 has a negative real part and a positive imaginary part and ξ_0' has a positive real part and a negative imaginary part. Combined with the factor of time $\exp(j\omega)$, $\exp\{j(\omega t - kx)\}$ gives the normal travelling wave, $\exp\{j(\omega t + \xi_0 x)\}$ gives the attenuating wave travelling in the positive x direction, and $\exp\{j(\omega t + \xi_0'x)\}$ gives the attenuating wave travelling in the negative x direction. This means that in zone II the first term is the normal wave, the second term is the entry-end-effect wave and the third term is the exit-end-effect wave; they have corresponding terms in the one-dimensional solution of Eq. (35).

9.5. PERFORMANCE CALCULATIONS

Having derived expressions for flux densities in the air gap, the methods of calculating performance of the linear induction motor will be explained.

For constant current drive the primary current is

$$j_1 = J_1 e^{j(\omega t - kx)} \quad \text{(A/m)} \tag{141}$$

and magnetic flux density in the air gap of zone II is given by

$$b_y = B_y \varepsilon^{j\omega t} . \tag{142}$$

Here B_y is given by $B_{3y}(x,y)$ in Eq. (130) or (139) with $y=a$. Then thrust generated between j_1 and b_y is given by

$$F = \int_0^L \mathrm{Re}[j_1]\,\mathrm{Re}[b_y]dx \qquad (\mathrm{N/m}) . \qquad (143)$$

For this calculation, the magnetic flux in zone II $(0 < x < L)$ is used, since outside of this zone there is no primary current and hence no thrust is generated. Equation (143) contains double-frequency components which generate noise but produce no time-average thrust. When the double-frequency components are neglected, thrust is given by

$$F = \frac{1}{2} \int_0^L \mathrm{Re}[j_1{}^* b_y]dx \qquad (\mathrm{N/m}), \qquad (144)$$

where $j_1{}^*$ is the conjugate of j_1. Thus we get from Eqs. (130) and (144)

$$F = \frac{\mu_0 J_1{}^2}{2}\,\mathrm{Re}\left[\frac{jkLG(-k,a)}{H(-k)}\right]$$

$$+ \frac{\mu_0 J_1{}^2}{2}\,\mathrm{Re}\left[\frac{\xi_0\{1 - \varepsilon^{j(\xi_0+k)L}\}\,G(\xi_0,a)}{(\xi_0+k)^2 H'(\xi_0)}\right]$$

$$+ \frac{\mu_0 J_1{}^2}{2}\,\mathrm{Re}\left[\frac{\xi_0{}'\varepsilon^{-j(\xi_0'+k)L}\{1 - \varepsilon^{j(\xi_0'+k)L}\}\,G(\xi_0',a)}{(\xi_0'+k)^2 H'(\xi_0')}\right] \qquad (\mathrm{N/m}) . \qquad (145)$$

Here the thrust is per unit width of one iron core in newton. The first term is the normal thrust, which will be obtained when there is no end effect. The second term is thrust component due to the entry-end-effect wave, and the third term is thrust component due to the exit-end-effect wave. As explained in Chapters 5 and 8, the thrust component due to the exit-end-effect wave is very small and can be neglected in most practical applications, and the thrust component due to the entry-end-effect wave has considerable influences on the thrust versus slip characteristics which are adverse for high-speed linear induction motors.

Now the secondary input will be calculated. Electric field intensity is given by $\mathbf{E} = -\partial \boldsymbol{\Phi}/\partial t$ and in the present case vector potential $\boldsymbol{\Phi}$ has a z component only and hence $\mathbf{E}$ has z component only. Using the expression of ϕ_3 given by Eq. (127), electric field intensity acting on the primary current is derived as follows:

$$E_{3z}(x,a) = -j\omega\mu_0 J_1 \left\{ \frac{G(-k,a)}{H(-k)}\exp(-jkx) + \frac{G(\xi_0,a)}{(\xi_0+k)H'(\xi_0)}\exp(j\xi_0 x) \right.$$

$$\left. + \frac{\exp\{-j(\xi_0'+k)L\}}{(\xi_0'+k)H'(\xi_0')}\,G(\xi_0'a)\exp(j\xi_0'x) \right\} . \qquad (146)$$

In a way similar to that of the thrust, neglecting the double frequency component, the secondary input per unit width of one iron core is given by

$$P_2 = \frac{1}{2} \int_0^L \mathrm{Re}[j_1{}^* e_z] dx .$$ (147)

Thus we get

$$P_2 = \frac{\omega\mu_0 J_1{}^2}{2} \mathrm{Re}\left[-\frac{jLG(-k,a)}{H(-k)} \right]$$

$$+ \frac{\omega\mu_0 J_1{}^2}{2} \mathrm{Re}\left[\frac{\{\varepsilon^{j(\xi_0+k)L} - 1\} G(\xi_0,a)}{(\xi_0+k)^2 H'(\xi_0)} \right]$$

$$+ \frac{\omega\mu_0 J_1{}^2}{2} \mathrm{Re}\left[\frac{\varepsilon^{-j(\xi_0'+k)L}\{\varepsilon^{j(\xi_0'+k)L} - 1\} G(\xi_0',a)}{(\xi_0'+k)^2 H'(\xi_0')} \right] .$$ (148)

Mechanical output per unit width of one iron core is given by

$$P_{m,0} = Fv .$$ (149)

Secondary efficiency is given by

$$\eta_2 = \frac{Fv}{P_2} .$$ (150)

Since performance calculation under constant current drive has been established performances calculation under constant voltage drive can be calculated along the same line as in Chapter 6.

9.6. SIMPLIFIED FORMULA FOR CALCULATING PERFORMANCE

When the secondary sheet is nonmagnetic and the air-gap length $2a$ is smaller, the following conditions are satisfied for the three poles in Fig. 35.

$$(a-b)\xi \ll 1, \qquad b\gamma \ll 1 .$$ (151)

Then Eq. (134) becomes a good approximation, and vector potentials in Eqs. (126)–(128) can be expressed in the following simpler forms:
In zone I $(x < 0)$

$$\phi_3(x,a) = \mu_0 J_1 \frac{G(\xi_0',a)[\exp\{-j(\xi_0'+k)L\} - 1]}{a(\xi_0'+k)(\xi_0'-\xi_0)} \exp(j\xi_0'x) .$$ (152)

In zone II $(0 < x < L)$

$$\phi_3(x,a) = \frac{\mu_0 J_1}{a}\left[\frac{G(-k,a)\exp(-jkx)}{(\xi_0+k)(\xi_0'+k)} \right.$$

$$+ \frac{G(\xi_0,a)\exp(j\xi_0 x)}{(\xi_0+k)(\xi_0-\xi_0')}$$

$$\left. + \frac{G(\xi_0',a)\exp\{-j(\xi_0'+k)L\}}{(\xi_0'+k)(\xi_0'-\xi_0)}\exp(j\xi_0'x) \right. . \tag{153}$$

In zone III $(L < x)$

$$\phi_3(x,a) = \frac{\mu_0 J_1}{a}\frac{G(\xi_0,a)[1-\exp\{-j(\xi_0+k)L\}]}{(\xi_0+k)(\xi_0-\xi_0')}\exp(j\xi_0 x) . \tag{154}$$

Corresponding to these vector potentials the y components of the magnetic flux density in the air gap are given by $B_{3y}(x,a) = -\partial\phi_3(x,a)/\partial x$ as follows:

In zone I $(x < 0)$

$$B_{3y}(x,a) = -\frac{j\mu_0\xi_0'J_1 G(\xi_0',a)[\exp\{-j(\xi_0'+k)L\}]}{a(\xi_0'+k)(\xi_0'-\xi_0)}\exp(j\xi_0'x) . \tag{155}$$

In zone II $(0 < x < L)$

$$B_{3y}(x,a) = \frac{j\mu_0 J_1}{a}\left[\frac{kG(-k,a)}{(\xi_0+k)(\xi_0'+k)}\exp(-jkx) \right.$$

$$- \frac{\xi_0 G(\xi_0,a)}{(_0+k)(\xi_0-\xi_0')}\exp(j\xi_0 x)$$

$$\left. - \frac{\xi_0' G(\xi_0',a)\exp\{-j(\xi_0'+k)L\}}{(\xi_0'+k)(\xi_0'-\xi_0)}\exp(j\xi_0'x) \right] . \tag{156}$$

In zone III $(L < x)$

$$B_{3y}(x,a) = \frac{j\mu_0\xi_0 J_1 G(\xi_0,a)[\exp\{-j(\xi_0+k)L\}-1]}{(\xi_0+k)(\xi_0-\xi_0')}\exp(j\xi_0 x) . \tag{157}$$

These are the simplified expressions for the vector potentials and the y components of the magnetic flux densities in the air gap. In many cases when inequality conditions (151) hold, the factors $G(\xi,a)$ are very near 1, and then Eqs. (152)–(157) are further simplified. Under this condition the expression of the thrust in Eq. (145) is simplified as follows:

$$F = \frac{\mu_0 J_1^2}{2a} \operatorname{Re}\left[\frac{jkL}{(\xi_0+k)(\xi_0'+k)}\right]$$

$$+ \frac{\mu_0 J_1^2}{2a} \operatorname{Re}\left[\frac{\xi_0[\exp\{j(\xi_0+k)L\}-1]}{(\xi_0+k)^2(\xi_0'-\xi_0)}\right]$$

$$+ \frac{\mu_0 J_1^2}{2a} \operatorname{Re}\left[\frac{\xi_0' \exp\{-j(\xi_0'+k)L\}[\exp\{j(\xi_0'+k)L\}-1]}{(\xi_0'+k)^2(\xi_0-\xi_0')}\right]. \tag{158}$$

And the expression of the secondary input in Eq. (148) is simplified as follows:

$$P_2 = \frac{\omega\mu_0 J_1^2}{2a} \operatorname{Re}\left[\frac{-jL}{(\xi_0+k)(\xi_0'+k)}\right]$$

$$+ \frac{\omega\mu_0 J_1^2}{2a} \operatorname{Re}\left[\frac{\exp\{j(\xi_0+k)L\}-1}{(\xi_0+k)^2(\xi_0-\xi_0')}\right]$$

$$+ \frac{\omega\mu_0 J_1^2}{2a} \operatorname{Re}\left[\frac{\exp\{-j(\xi_0'+k)L\}[\exp\{j(\xi_0'+k)L\}-1]}{(\xi_0'+k)^2(\xi_0'-\xi_0)}\right]. \tag{159}$$

Chapter 10

Calculations on the Magnetic Field in the Air Gap*

10.1. NATURE OF POLES OF FOURIER TRANSFORMS

Poles of Fourier transforms of vector potentials play very important roles in the determination of travelling magnetic waves in the air gap. One of them is $\xi = -k$ of Eq. (120), and others are roots of Eqs. (121) and (122), of which approximate roots are given by Eq. (123). For the non-magnetic secondary Eq. (123) is written below.

$$H(\xi) = a\left(\xi^2 + j\frac{b}{a}v\mu_0\sigma_2\xi + j\frac{b}{a}\omega\mu_0\sigma_2\right) = a(\xi - \xi_0)(\xi - \xi_0') . \tag{160}$$

Starting with values in Eq. (160), accurate values of the two roots can be obtained with a computer.

It may be interesting to note that the approximate roots ξ_0 and ξ_0' given by Eq. (160) are identical to the characteristic roots obtained in the one-dimensional analysis in Chapter 4, except for factor j. ξ_0 and ξ_0' are complex numbers and can be written as below.

$$\xi_0 = -\beta - j\gamma_1 , \qquad \xi_0' = \beta - j\gamma_2 . \tag{161}$$

Then from Eq. (160) we get

$$\beta = \frac{\sqrt{2}\,\omega\mu_0\sigma_2'}{\{(v\mu_0\sigma_2')^2 + \sqrt{(v\mu_0\sigma_2')^4 + 16(\omega\mu_0\sigma_2')^2}\}^{1/2}} , \tag{162}$$

$$\gamma_1 = \frac{1}{2}\left(v\mu_0\sigma_2' - \frac{\omega\mu_0\sigma_2'}{\beta}\right) , \tag{163}$$

$$\gamma_2 = \frac{1}{2}\left(v\mu_0\sigma_2' - \frac{\omega\mu_0\sigma_2'}{\beta}\right) , \tag{164}$$

$$\sigma_2' = \frac{b}{a}\sigma_2 . \tag{165}$$

* Reported originally in refs. 6, 7 and 14.

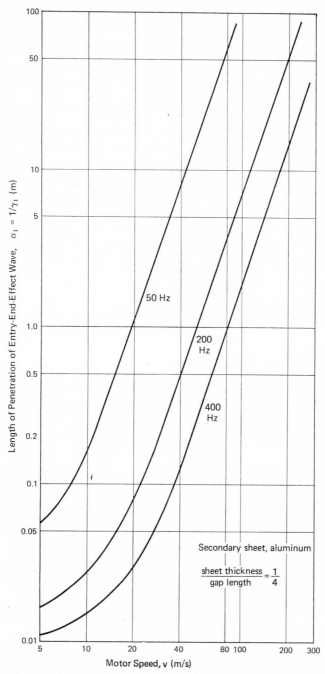

Fig. 36. Length of penetration of entry-end-effect wave versus motor speed.

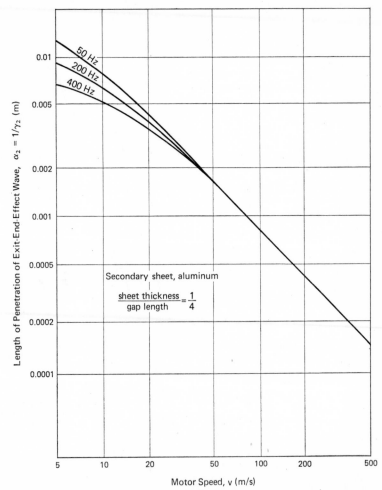

Fig. 37. Length of penetration of exit-end-effect wave versus motor speed.

Comparing ξ_0 and ξ_0' with characteristic roots of the one-dimensional analysis in Eqs. (23) and (24), the following relation should hold

$$j\xi_0 = k_1 ,$$
$$j\xi_0' = k_2 , \tag{166}$$

whence we get

$$\left.\begin{array}{l} \beta = \dfrac{\pi}{\tau_e}\ , \\[2mm] \gamma_1 = \dfrac{1}{\alpha_1}\ , \\[2mm] \gamma_2 = \dfrac{1}{\alpha_2}\ . \end{array}\right\} \qquad (167)$$

It is not difficult to prove these equations.

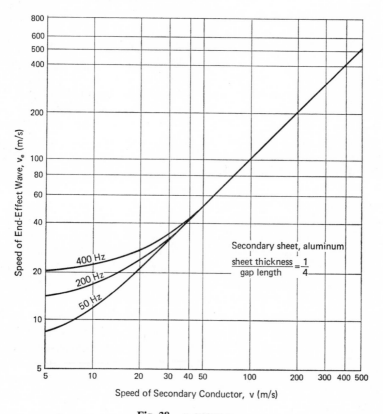

Fig. 38. v_e versus v.

$1/\gamma_1$ and $1/\gamma_2$ are the lengths of penetration for the entry-end-effect wave and the exit-end-effect wave, respectively, and $\tau_e = \pi/\beta$ is the half-wave length of the end-effect waves. Exact values for these lengths, which are calculated by the two-dimensional analysis, should be slightly different from those values in Eq. (167). Some of these values calculated with the two-dimensional analysis are shown in Figs. 36 and 37. Figure 38 is the exact values of the speed v_e of the end-effect waves calculated by the two-dimensional analysis by the following formula.

$$v_e = \frac{2\pi f}{\beta}.\qquad\qquad(168)$$

10.2. Calculation of the Magnetic Flux Density Distribution in the Air Gap and the Adequacy of the Two-dimensional Model

In the two-dimensional analysis of the linear induction motor explained in Chapter 9, it is assumed that the primary iron cores extend in the x direction and the y direction, as shown in Fig. 34. The end effect is caused only by the primary current, which is limited in length, and it is possible to assume that the calculated end effect is quite different from the actual one, and the calculated flux-density distribution and the calculated motor performance based on the two-dimensional model involve considerable error. However, this is not true. The two-dimensional analysis in Chapter 9 produces calculated results which are in good agreement with measured results. In this section magnetic flux density will be calculated and compared with measured results, and an explanation of why the assumption of the extended primary iron cores is correct will be given.

The two-dimensional analysis gives the y component of magnetic flux density by Eqs. (129)–(131) or (138)–(140). Inserting $y=a$ in these equations, the y component B_{3y} on the surface of the primary iron cores is obtained. Only the y component B_{3y} produces thrust which can be measured with search coils placed on the surface of the primary iron cores. The resultant maximum value of B_{3y} as a function of x was calculated using Eqs. (138)–(140), which are valid for the nonmagnetic secondary sheet. An example of such a calculation is shown in Fig. 39. The flux density distribution curves in the figure indicate clearly influence of the end effect which weakens the magnetic field near the entry end. In Figs. 50 and 51, calculated flux-density-distribution curves are shown together with the measured curves. The calculated curves are in good agreement with the measured curves, except for some irregularity of the latter curves, which is caused by discrete current distribution in slots of the actual primary winding.

Although the agreement is good within the air-gap zone II $(0 < x < L)$, there is discrepancy in zone III $(L < x)$. In Fig. 39 the distribution curves decay rather slowly in zone III $(L < x)$ and there is considerable magnetic

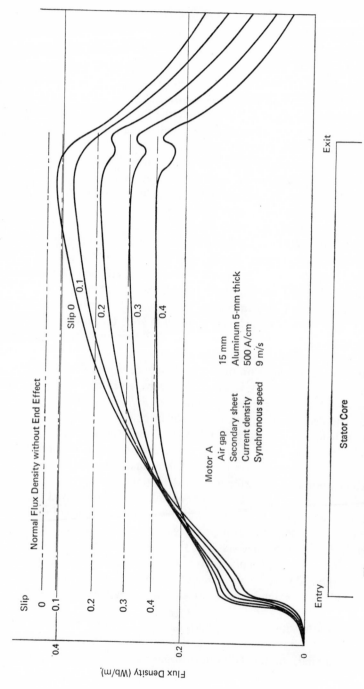

Fig. 39. Magnetic-flux-density-distribution curves calculated (on the primary core surface).

flux in zone III, which is given by Eq. (140). Actually there is no magnetic flux in zone III except for fringing magnetic flux, which is very slight. The discrepancy is caused by the artificially assumed extension of the iron core in zone III, as shown in Fig. 34. At the exit end most of the forward travelling wave in zone II is transmitted into zone III and only a small part of it is reflected. Accordingly the third term of Eq. (139), representing the reflected wave, is much smaller than the actual wave, and the transmitted wave given by Eq. (140) is much larger than the actual wave, which is practically null. This rather large discrepancy of the magnetic flux density distribution near the exit end involves very small errors in the performance calculations of the linear induction motor, which can be neglected in most practical performance calculations. As was explained in Chapter 5, the reflected wave decays so quick that its existence is limited to a small vicinity very close to the exit end and its influence on motor performance is negligible. Accordingly the discrepancy relative to the reflected wave does not introduce any error into performance calculation.

The transmitted wave in zone III, which is much larger than the actual wave, does not also introduce any error into performance calculations because there is no primary winding in zone III on which the transmitted wave acts to produce thrust and electromotive force. The transmitted wave does not generate more than a small amount of secondary copper loss; its influence on motor performance is also negligible.

In zone I $(x<0)$ there is a travelling wave given by Eq. (138). This is the backward travelling wave which decays very quickly due to factor exp $(j\xi_0'x)$. The decay is so quick that it can be considered that practically no magnetic field exists in zone I, in spite of the existence of an extended primary iron cores there. It can be stated that the artificially assumed extension of the primary cores in zone I introduces practically no errors in field calculation and in performance calculation.

Chapter 12 will show how other calculated and measured characteristic curves are in good agreement, which is another evidence of the validity of the assumed extension of the primary iron cores of the two dimensional model. It should be pointed out here that the transmitted wave in zone III given in Eq. (128) or (137) plays a very important role in the compensation theory of the end effect. This is one of the reasons why it is assumed that the primary cores extend beyond longitudinal length of the air gap.

10.3. Two-dimensional Field Distribution without the End Effect

Vector potentials of the two-dimensional field of the linear induction motor were obtained as Eqs. (126), (127) and (128). In these vector potentials terms with $\exp(j\xi_0 x)$ or $\exp(j\xi_0' x)$ are the end-effect waves and a term with $\exp(-jkx)$ is the normal wave. It is interesting to find magnetic field distribution when there is no end effect, although such field distribution can exist only in the air gap of a rotating induction motor or in a region of the air gap far from both entry and exit ends of a very long linear induction motor. When there is no end effect, all the terms of Eqs. (126), (127) and (128), except the first term in region 2 which is written below, disappear.

$$\phi_{3n}(x,y) = \mu_0 J_1 \frac{G(-k,y)}{H(-k)} \varepsilon^{-jkx} . \tag{169}$$

This is the vector potential of the normal wave in the air gap, whose permeability is $\mu_3 = \mu_0$. Vector potential of the normal wave in the secondary sheet can be obtained from Eq. (116), taking the term for $\xi = -k$ and making L infinity. Thus we get

$$\phi_{3n}(x,y) = \mu_0 \frac{\cosh \gamma_k y}{H(-k)} \varepsilon^{-jkx} , \tag{170}$$

where γ_k is γ of Eq. (122) with $\xi = -k$. With these vector potentials magnetic flux densities are derived from Eq. (80) as follows:
In region 3 (the air gap)

$$B_{3x}(x,y) = k\mu_0 J_1 \frac{1}{H(-k)} \Big\{ \cosh b\gamma_k \sinh(y-b)k$$

$$+ \frac{\mu_0}{\mu_2} \frac{\gamma_k}{k} \sinh b\gamma_k \cosh(y-b)k \Big\} \varepsilon^{-jkx} , \tag{171}$$

$$B_{3y}(x,y) = -jk\mu_0 J_1 \frac{1}{H(-k)} \Big\{ \cosh b\gamma_k \cosh(y-b)k$$

$$+ \frac{\mu_0}{\mu_2} \frac{\gamma_k}{k} \sinh b\gamma_k \sinh(y-b)k \Big\} \varepsilon^{-jkx} . \tag{172}$$

In region 2 (the secondary sheet)

$$B_{2x} = \gamma_k \mu_0 J_1 \frac{\sinh \gamma_k y}{H(k)} \varepsilon^{-jkx} , \tag{173}$$

$$B_{2y}(x,y) = -jk\mu_0 J_1 \frac{\cosh \gamma_k y}{H(-k)} \varepsilon^{-jkx} . \tag{174}$$

Here $H(-k)$ is given from Eq. (119) with $\xi = -k$.

S. Nonaka and K. Yoshida[18] derived expressions similar to Eqs.

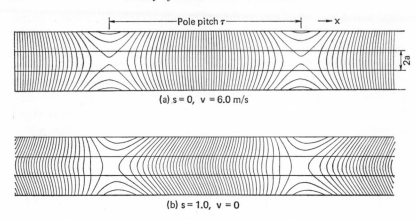

Fig. 40.   Flux pattern in air gap for copper secondary.

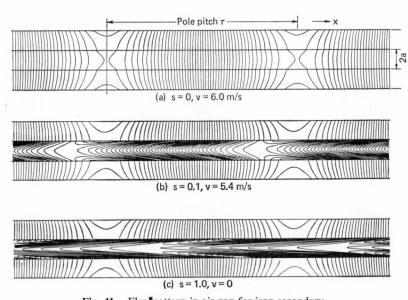

Fig. 41.   Flux pattern in air gap for iron secondary.

(171)–(174). Their analysis does not take into account the end effect. However, as far as the normal wave is concerned, their expressions and our expressions are the same. Although both expressions appear to be different, it is not difficult to check the identity of each. Nonaka and Yoshida drew excellent field mappings and Figs. 40 and 41 were taken from their article.[18] Figure 40 illustrates the case of a copper secondary sheet 5-mm thick. The primary frequency is 60 Hz and the synchronous speed is 6 m/s. In the case of the copper secondary, the skin effect is not recognizable even for slip. 1. For slip 0, 90% of the total flux leaving one iron core reaches the other iron core, and for slip 1, 84% reaches the other side. In the case of the iron secondary shown in Fig. 41, the skin effect is very clearly recognizable except for slip 0. For slip 0, 94% of the total flux reaches the other side of the iron core, and for slip 1 only 12% reaches the other side. In these examples the ratio between the pole pitch and the total gap length ($2b$) is 50 mm : 15 mm. This ratio is rather small. In high-speed linear induction motors this is much larger, therefore, the percentage of the crossing magnetic flux is also much larger. Current distribution in the copper secondary is rather uniform in the y direction and under this condition the one-dimensional model, where the x component of the magnetic flux is neglected, is a good approximation. On the other hand, in the iron secondary the x component of the magnetic flux is very large, and the magnetic flux and current are concentrated near the surfaces of the secondary sheet due to the skin effect. The one-dimensional model is then a poor approximation. It is anticipated that whether the one-dimensional model is a good approximation or not depends on whether or not the x component of the magnetic flux is large. For a good approximation of the one-dimensional model the x component of the magnetic flux must be small both in the air gap and in the secondary sheet. This means that ratio between the net gap length and pole pitch $(a-b)/\tau$ must be small and the skin effect must not be prominent in the secondary.

The condition in which the x components are smaller can be qualitatively determined by means of Eqs. (171)–(174). From Eq. (171) the conditions necessary for B_{3x} to be small are that $\sinh(y-b)k$ and $\sinh b\gamma_k$ are small. These conditions are satisfied if the following relations hold:

$$|b\gamma_k| \ll 1 , \qquad |(a-b)k| \ll 1 . \tag{175}$$

And the above-mentioned condition must also hold, that is

$$\frac{a-b}{\tau} \ll 1 . \tag{176}$$

When these conditions hold, B_{2x} of Eq. (173) is much smaller than B_{2y} of Eq. (174), B_{3x} of Eq. (173) is much smaller than B_{3y} of Eq. (174) and the x components can be neglected. These are the approximations that were done in the one-dimensional expressions of the magnetic field in Chapter 4. γ_k of inequality (175) is given by Eq. (122), and can be transformed as follows

$$\gamma_k = k\sqrt{1 + j\frac{\mu_2\sigma_2}{k}(v_s - v)}$$

$$= k\sqrt{1 + j\frac{\mu_2\sigma_2 s\omega}{k^2}} ,$$

$$s = \frac{v_s - v}{v_s} . \tag{177}$$

For the aluminum secondary we have $\mu_2 = 4\pi \times 10^{-7}$, $\sigma_2 = 1/2.8 \times 10^{-8}$, and then for $\omega = 2\pi f = 100\pi$, $k = \pi/\tau = 3.14$, $s = 0.05$ (synchronous speed $v_s = 100$ m/s), we get

$$\frac{1}{k^2}\mu_2\sigma_2 s\omega = 70 ,$$

which is much larger than 1. And hence Eq. (177) can be approximated by

$$\gamma_k = \sqrt{\frac{\mu_2\sigma_2 s\omega}{2}} + j\sqrt{\frac{\mu_2\sigma_2 s\omega}{2}} . \tag{178}$$

For the above example $\gamma_k = 6 + j6$. For $a = 0.02$ and $b = 0.01$, $b\gamma_k = 0.06 + j0.06$, $(a-b)k = 0.032$ and $(a-b)/\tau = 0.01$. Then the inequalities (175) and (176) hold.

When the secondary sheet is iron, μ_2 is much larger. For $\mu_2 = 500 \times 4\pi \times 10^{-7}$, we get $\gamma_k = 135 + j135$, $b\gamma_k = 1.35 + j1.35$ and $(a-b)k = 0.032$. Then $|b\gamma_k| \ll 1$ does not hold and this means that the one-dimensional field is not valid even approximately for the iron secondary.

Skin depth is given from Eq. (178) as

$$\delta = \sqrt{\frac{2}{\mu_2\sigma_2 s\omega}} . \tag{179}$$

In the above examples $\delta = 0.16$ m for copper and $\delta = 0.0073$ m for iron. For copper the skin depth is much larger than thickness of the secondary sheet. Whether the skin depth is much larger than thickness of the secondary sheet or not is a good criterion for determining whether or not the one-dimensional expressions of the magnetic field are valid. It should be also mentioned that when skin depth is smaller than thickness of the secondary sheet good performance of the linear induction motor can not be expected. Since the field and also current can only penetrate a very thin layer near the surface of the secondary sheet, most of the secondary sheet remains inactive and each side of the two-sided linear induction motor acts rather independently of each other.

Another problem is the independence of the primary iron cores on each side of the machine with an iron secondary, which results in a large lateral pull because the total magnetic flux in each of the air gaps is different if air gaps on each side of the secondary sheet are different from each other.

In this chapter attention has been paid only to the normal wave, however, the same argument holds also for the end-effect waves. Replacing $-k$ with the other two poles ξ and ξ_0', the same criterion given by Eq. (175) is also applicable to the end-effect waves. Since $k > |\xi_0|, |\xi_0'|$ in many cases, the criterion is met for the nonmagnetic secondary and the one-dimensional approximation is also good for the end-effect waves.

Chapter 11

Calculation of Performance by Means of the Two-dimensional Solution*

After deriving the solutions for the magnetic field in the air gap, motor performance can be determined according to the procedures explained in Section 9.5. The solutions for air gap field, which will be utilized in performance calculation, are the two-dimensional solutions obtained in Chapter 9.

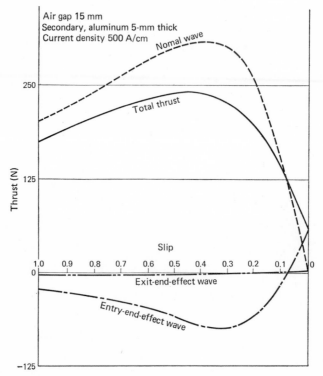

Fig. 42. Thrust-versus-slip curves for motor A with aluminum secondary.

* Reported originally in refs. 14, 16, 23 and 27.

First, the performance of motor A in the table in Appendix VI under constant current drive was calculated. The thrust-versus-slip curve is given in Fig. 42. The curve marked "normal wave" indicates thrust without the end effect. The curve marked "entry-end-effect wave" indicates thrust generated by the entry-end-effect wave, which is negative thrust except in small slip region. The curve marked "exit-end-effect wave" indicates thrust generated by the exit-end-effect wave. This thrust is very small and can be neglected, and this is experimental proof that the influence of the exit-end-effect wave is very small and can be neglected. The curve marked "total thrust" indicates the resultant thrust, which is the algebraic sum of the three component thrusts. The total thrust is smaller than the normal thrust in the large slip region and is larger than the normal thrust in the small slip region. Thrust is still generated even at synchronous speed. It can be seen that the positive thrust at synchronous speed comes from the

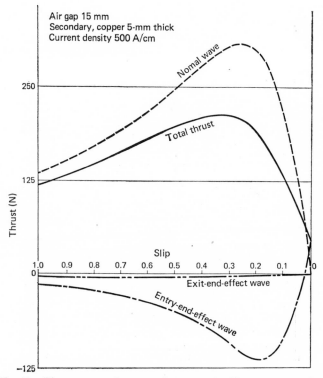

Fig. 43. Thrust-versus-slip curves for motor A with copper secondary.

entry-end-effect wave. As explained previously in Chapter 7, this characteristic is typical of the low-speed motor. This motor has synchronous speed
of 9 m/s, which is low speed.

Figure 43 shows similar curves for the motor A whose secondary sheet
has been replaced with a copper sheet. The general trend is the same as in
Fig. 42. Figure 44 shows the thrust-versus-slip curves for different gap
lengths; as the gap length increases, the total thrust decreases. Figure 45
shows the thrust-versus-slip curves indicating the influence of secondary
resistance. Maximum values of thrust without the end effect are the same
for aluminum secondary and copper secondary, although slips at which
maximum thrusts occur are different. This means that principle of proportional shifting holds for the thrust of the linear induction motor when no
end effect exists. When the end effect occurs, maximum thrust is different

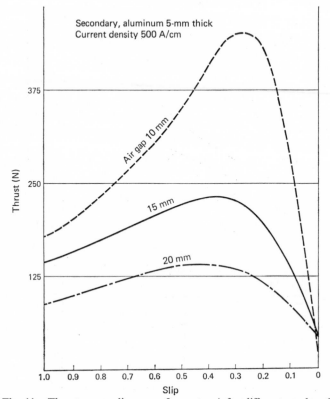

Fig. 44.   Thrust-versus-slip curves for motor A for different gap length.

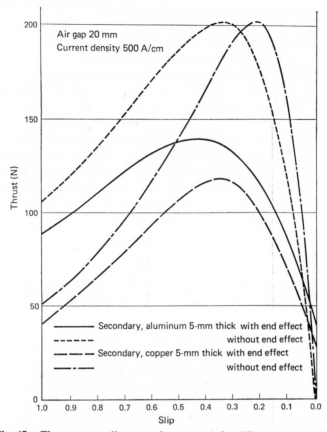

Fig. 45. Thrust-versus-slip curves for motor A for different secondaries.

for the copper secondary and the aluminum secondary; that is, the principle of proportional shifting does not hold when the end effect occurs.

Some examples of the calculated performance of a high-speed motor under constant voltage drive are shown here. The motor under investigation is motor D in Appendix VI, whose synchronous speed is 150 m/s.

Figure 46 shows calculated thrust-versus-slip curves, and Fig. 47 shows the calculated power factor-versus slip and efficiency-versus-slip curves. Curves marked "without end effect" show calculated values which do not take the end effect into account, and curves marked "with end effect" show calculated values which take end effect into account. Calculations are made for motors running under constant voltage drive in the

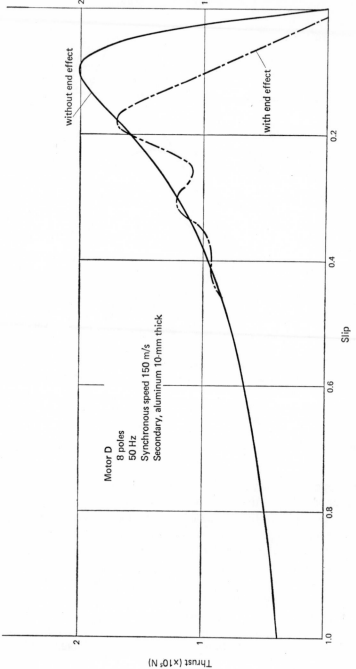

Fig. 46. Thrust-versus-slip curves for motor D under constant voltage drive.

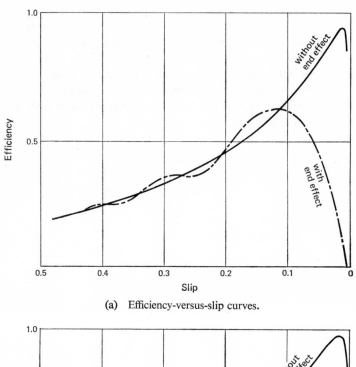

(a)   Efficiency-versus-slip curves.

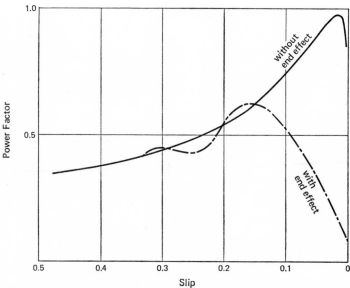

(b)   Power factor-versus-slip curve.

Fig. 47.   Characteristic curves for motor D under constant voltage drive.

same way as explained in Chapter 6. In large slip region the end effect has little influence on performance curves, while in the small slip region it has considerable influence. The end effect reduces thrust, power factor and efficiency to a large extent in the small slip region, which is the important running region. All these influences of the end effect are so adverse that the feasibility of the linear induction motor for high-speed applications may be lost if they are not eliminated, and some remedy may be necessary for realization of such a motor.

Chapter 12

Experimental Results*

Experimental results on the performance of the linear induction motor are very few, especially in respect to the high-speed linear induction motor. Experiments on the linear induction motor involve some difficulty because the dimensional needs of the linear motor are large compared with those of the rotating motor. This increases the cost of experimental devices and makes measurements more difficult. In order to avoid these difficulties the rotating disc-type induction motor was contrived for experimental purposes in order to have an induction motor equivalent in function to a linear induction motor. In this type of motor a rotating con-

Fig. 48.   Motor E in Appendix VI.

* Reported originally in refs. 6, 9, 14, 23 and 27.

ductive metal disc takes place of the secondary conductive sheet of the
linear induction motor. Two-sided primary iron cores are placed along the
periphery of the disc to drive it. Peripheral speed is different at different
radial distances from the center of the disc, however, if speed and slip are
determined at the middle point of the iron stack, there will be no noticeable
error in performance calculations based on slip thus determined. As long
as material and thickness of the disc are, respectively, the same as those
of the secondary sheet of the linear induction motor, and the gap length
and primary sides are the same in both motors, the rotating disc motor
can be considered equivalent to the linear induction motor.

Motors A and E shown in Appendix VI were actually built for our
experimental purposes. Figure 48 shows motor E, which is a rotating-disc-
type motor. Motor A is also a disc-secondary-type linear induction motor,
which has an aluminum disc and 2-m wide in diameter, and two-sided
primary iron cores. There are three sets of the iron cores which have
three-phase windings that are connected as shown in Fig. 23 in order to
obtain balanced three-phase currents in spite of the unbalancing influence
of the end effect. The primary windings are single-layer windings as shown
in Fig. 2, and therefore create uniform ampere turns over the entire length
of the iron core. Magnetic flux density distribution was measured by
search coils wound around each tooth. Voltage at the terminals of the
search coils provides flux density and Fig. 49 shows the flux density
distribution thus obtained; the numbers along the absissa are tooth
numbers. At the entry end the magnetic field is weakened by the entry-
end effect; the field is built up gradually and there are dips and sharp
rises on the distribution curves near the exit end. These curves were measured for constant current drive; that is, all the curves were measured with
the same current. A distribution curve (marked "without disc") for the
case when the rotating disc was removed is also shown in the figure. In
this case there is no secondary current and the curve should give the distribution curve for synchronous speed (zero slip) without the end effect.
It should be noted that the measured distribution curves for smaller slip
reach higher levels than the distribution curve for motors without discs,
although they start from much lower level at the entry end. Figure 50
shows the measured flux-density-distribution curves having the same tendencies as those in Fig. 49, for motor A with a rotating disc 5-mm thick.
The flux density distribution curves calculated by the method described
in Section 10.2 are also drawn as broken lines. The measured curves have
irregularities due to discrete distribution of the ampere turns in slots and
the rise caused by the reflected wave near the exit wave. Although these

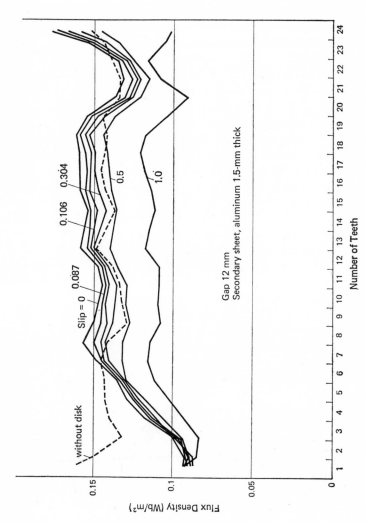

Fig. 49. Measured magnetic-flux-density-distribution curves for motor A.

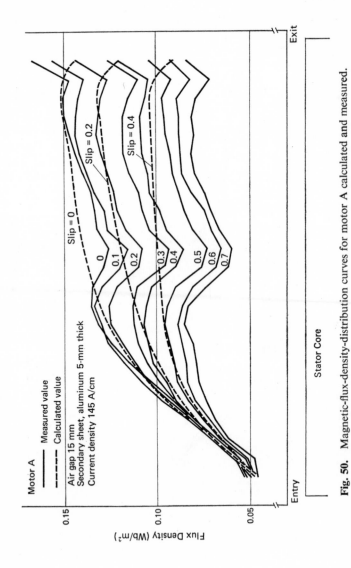

Fig. 50. Magnetic-flux-density-distribution curves for motor A calculated and measured.

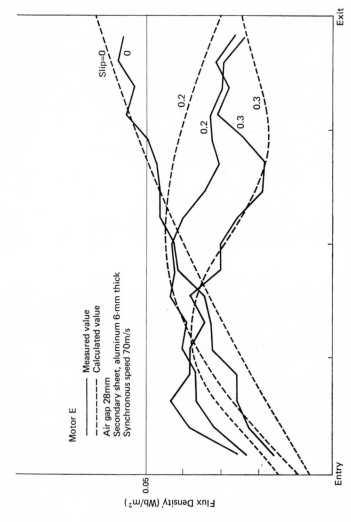

Fig. 51. Magnetic-flux-density-distribution curves for motor E calculated and measured.

irregularities are missing in the calculated curves, agreement between the measured and calculated curves is considered to be good.

Figure 51 shows the measured magnetic-flux-density-distribution curves for motor E in the table of Appendix VI. The flux-density-distribution curves calculated by the two-dimensional solution in Chapter 9 are also shown as the broken lines in the figure. Both measured and calculated curves are in good agreement. It should be noted that the curve for slip of 0.3 clearly indicates undulation, which is characteristic of the high speed motor.

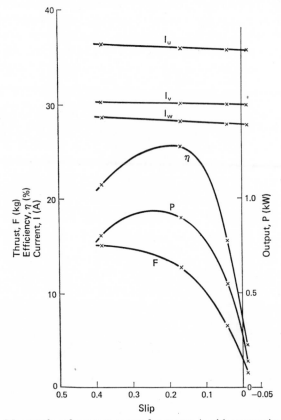

Fig. 52. Measured performance curves for motor A with conventional winding.

Figure 52 shows measured results of the primary current, thrust, output, the power factor and efficiency of motor A, which was driven with a constant three-phase voltage source; the abssissa indicates slip. Only one set of the three sets of iron cores shown in Fig. 23 was used for this measurement. Hence, primary currents are considerably unbalanced due to the end effect as was explained in Chapter 6; current of phase *u*, which is located nearest to the entry end, is the largest. Figure 53 shows measured performance curves obtained, when all of the three sets of the primary iron cores were used and three sets of the primary windings were connected as explained in Chapter 6, all values referring to one set of the iron core. In this case the primary currents are well balanced, and hence performance is improved compared with that of the unbalanced motor shown in Fig. 52.

Both the power factor and efficiency are very low. This is due to the large air gap. It should be pointed out that larger air gap degrades the performance of the low-speed linear induction motor but it does not necessarily degrade the performance of the high-speed linear induction motor, as was explained in Chapter 7. As was also explained in Chapter 7, the low-speed motor produces thrust at and above synchronous speed due to

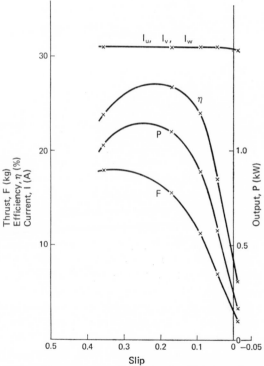

Fig. 53. Measured performance curves for motor A with balanced-current connection.

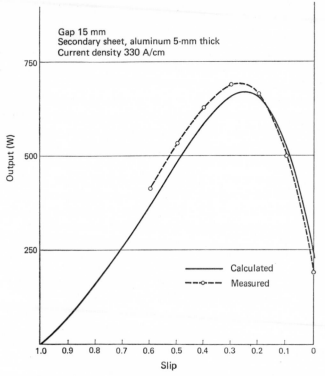

Fig. 54. Measured and calculated output-versus-slip curves for motor A.

the end effect. This is clearly recognized in Figs. 52 and 53. Figure 54 shows measured and calculated output-versus-slip curves for motor A; agreement is good in the small slip region. It should be pointed out also, that the primary currents are almost constant over a wider slip range in Figs. 52 and 53, although all the measurements were made under constant voltage drive. In ordinary induction motors the primary current decreases as slip approaches zero, and at zero slip only exciting current flows. However, the end effect keeps the primary current at a high level even in the small slip region. Besides this, the exciting current of the low-speed induction motor is larger, resulting in poorer overall performance.

Motor E in the table of Appendix VI was tested as an example of a high-speed linear induction motor (synchronous speed 70 m/s), and the results are shown in Fig. 55(a), (b), (c) and (d). Performance calculation was also carried out by the Fourier transform method described in

Chapter 9. The results of performance calculation are also shown in Fig. 55(a), (b), (c) and (d); agreement between the measured and calculated curves is good. It should be pointed out that this would be the first time that the degradation of performance of a high-speed linear induction motor due to the end effect was proved experimentally and the first time that performance calculation of the high-speed linear induction motor, which takes the end effect into account, was confirmed experimentally. As these figures clearly indicate, the performance degradation due to the end effect is severe and serious. The performance curves were measured or calculated for motor operations under constant voltage drive. The current drawn by the linear induction motor in the small slip region, which is the important motor run region, is much larger than in the case where there is no end effect, as Fig. 55 (d) indicates, and the primary winging may then be overheated. If it is assumed that the same current flows and the same copper loss is generated in the primary winding, then the reduction of the thrust due to the end effect becomes much more drastic. The

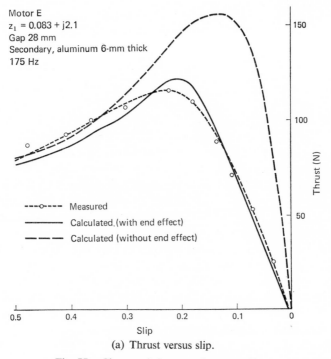

(a) Thrust versus slip.

Fig. 55. Characteristic curves for motor E.

example shown in Fig. 55 clearly indicates that some means must be taken to eliminate the end effect, in order to reestablish the feasibility of a high-speed linear induction motor, which was lost as a result of the presence of the end effect. Steps to eliminate the end effect will be shown in the following chapters.

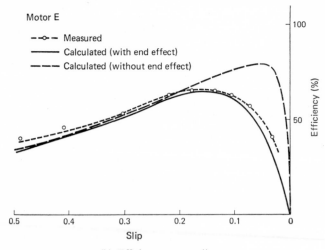

(b) Efficiency versus slip.

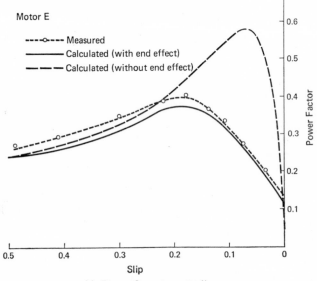

(c) Power factor versus slip.

Fig. 55. Continued.

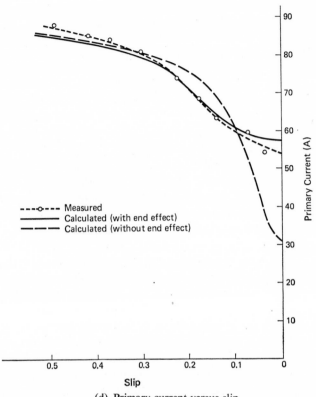

(d) Primary current versus slip

Fig. 55. Continued

Chapter 13

Compensation of the End Effect*

13.1. INTRODUCTION

In the preceding chapters it was shown both theoretically and experimentally that the end effect exercises considerable influence on linear induction motor performance. Influence of the end effect is quite different for low-speed and high-speed motors. In low-speed motors the influence is relatively small and even advantageous, while in high-speed motors the influence is large and disadvantageous. A large amount of experimental data has been available for low-speed motors, while practically none has been available for high-speed motors. Because of the lack of experimental data on high-speed motors and the incompleteness of theoretical analysis, the seriousness of the adverse influences of the end effect were overlooked and underestimated. As was clarified in preceding chapters, these influences are so seriously adverse in high-speed linear induction motors that its feasibility might be lost for most applications if certain countermeasures are not taken into consideration. As explained in Chapter 8, the use of a larger number of poles is to a limited extent effective for alleviating the adverse influence, although it does not eliminate it. Selection of proper parameters, such as the secondary conductivity, the gap length, and the power source frequency, etc., may also help to alleviate the end-effect influence. Our theory of the linear induction motor made it possible to calculate the linear induction motor performance under the influence of the end effect and revealed that the effectiveness of the proper selection of parameters is rather limited.

Our laboratory succeeded in establishing the theory of compensation of the end effect. Based on the compensation theory, we contrived a compensated linear induction motor, in which a compensating winding eliminates the end effect at the entry end of the main winding. The compensation theory and the compensated linear induction motor will be explained below.

* Reported originally in refs. 15, 16, 22, 23 and 27.

13.2. A Compensated Linear Induction Motor
of the Two-windings Type

In the linear induction motor the end effect generates two end-effect waves in the air gap; one is the entry-end-effect wave, which travels in the same direction as the normal travelling wave and the other is the exit-end-effect wave, which travels in the opposite direction. Both end-effect-waves attenuate while travelling. The exit-end-effect wave attenuates so quickly that it has practically no influence on the motor performance and can be neglected as was explained previously. The object of the compensation is then to eliminate the entry-end-effect wave.

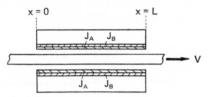

Fig. 56. Compensated linear induction motor of the two-windings type.

A schematic picture of the compensated linear induction motor of the two-windings type is shown in Fig. 56. This type of motor has two primary windings which are placed along-side with each other to act on the common secondary sheet. The number of poles P_A of winding A is different from the number of poles P_B of winding B. The length of the windings and of the primary iron cores are the same and given by L. In the model in Fig. 34, on which the derivation of Eqs. (126)–(128) was based, the primary iron cores were unlimited. The unlimited extension of the core length introduces very little error in the entry-end effect.

Let currents of winding A and winding B be expressed by J_A and J_B, respectively. Then Eq. (127) gives vector potentials produced by winding A and winding B, respectively, as follows:

$$\phi_{3A}(x,y) = \mu_3 J_A \left\{ \frac{G(-k_A,y)}{H(-k_A)} \exp(-jk_A x) \right.$$

$$\left. + \frac{G(\xi_0,y)}{(\xi_0+k_A)H'(\xi_0)} \exp(j\xi_0 x) \right\}, \tag{180}$$

$$\phi_{3B}(x,y) = \mu_3 J_B \left\{ \frac{G(-k_B,y)}{H(-k_B)} \exp(-k_B x) \right.$$

$$\left. + \frac{G(\xi_0,y)}{(\xi_0 + k_B)H'(\xi_0)} \exp(j\xi_0 x) \right\} . \tag{181}$$

Here the unimportant exit-end-effect waves are neglected. The travelling constants k_A and k_B in these equations are given, respectively, by

$$\left. \begin{array}{l} k_A = \dfrac{\pi}{\tau_A} = \dfrac{P_A \pi}{L} , \\[2ex] k_B = \dfrac{\pi}{\tau_B} = \dfrac{P_B \pi}{L} . \end{array} \right\} \tag{182}$$

Vector potentials $\phi_{3A}(x,y)$ and $\phi_{3B}(x,y)$ are superposed in the air gap, and, hence, the resultant vector potential for the entry-end-effect wave is given by the sum of the second terms of Eqs. (180) and (181). Thus we get

$$\phi_{3\text{end-effect}} = \frac{\mu_3 G(\xi_0,y)}{H'(\xi_0)} \exp(j\xi_0 x) \left(\frac{J_A}{\xi_0 + k_A} + \frac{J_B}{\xi_0 + k_B} \right) . \tag{183}$$

If the bracketed part of Eq. (183) is zero, the resultant vector potential of the entry-end-effect wave is zero everywhere in the air gap. Thus the condition of eliminating the end-effect wave is given by the following equation.

$$\frac{J_A}{\xi_0 + k_A} = - \frac{J_B}{\xi_0 + k_B} . \tag{184}$$

This is the compensation condition to be satisfied for the elimination of the end effect. When this condition is satisfied, the vector potential in the air gap is the sum of the first terms of Eqs. (180) and (181), which are both normal waves. Thus we have

$$\phi_3(x,y) = \mu_3 \left\{ \frac{J_A G(-k_A,y)}{H(-k_A)} \exp(-jk_A x) \right.$$

$$\left. + \frac{J_B G(-k_B,y)}{H(-k_B)} \exp(-jk_B x) \right\} . \tag{185}$$

If both numbers of poles P_A and P_B are equal, from Eq. (182) we have $k_A = k_B$. Then Eq. (184) gives $J_A = -J_B$ and $\phi_3(x,y)$ of Eq. (185) is zero. Thus $P_A = P_B$ gives a trivial solution and therefore P_A must be different from P_B.

In high-speed motors Eq. (39) holds, and ξ_0 is approximated by

$$\xi_0 \doteq -\frac{\pi}{\tau_e}$$

$$= -\frac{\pi}{\tau(1-s)} \ . \tag{186}$$

If slip s refers to winding A, then from Eqs. (182), (184) and (186), we get

$$J_B \doteq \frac{P_B - P_A}{s P_A} J_A \ . \tag{187}$$

For a given slip the smallest J_B is obtained when the difference of numbers of poles is given by

$$|P_B - P_A| = 2 \ . \tag{188}$$

It should be pointed out that the two windings act independently of each other in producing thrust, when the difference of numbers of poles is two. In this respect they are two independent linear induction motors acting on a common secondary sheet

It should be also pointed out that the compensated linear induction motor of the two-windings type is an unecnnomical motor. For example, for $s = 0.08$, $P_A = 10$ and $P_B = 8$, we get $J_B = 2.5 J_A$. Thus current carrying capacity of winding B must be 2.5 times larger than that of winding A. The compensated linear induction motor of this type is much more uneconomical than the compensated linear induction motor of the compensating-winding type which will be explained in the next section.

13.3. A Compensated Linear Induction Motor of the Compensating-winding Type

The compensated linear induction motor of the two-windings type in Section 13.2 is not practical because it has two large primary windings. Both of them are equal in length and cover the entire length of the iron core. At any motor speed, except zero, slips are different for the two windings because their pole pitches are different. In small slip region one winding with smaller slip produce more power than the another with larger slip. The former draws less current than the latter. This situation leads to an uneconomically compensated linear motor so that a more economically compensated one will be developed.

Figure 57 shows a schematical picture of a more economically compensated linear induction motor. It has two primary windings; one is the main winding and is located between $x = 0$ and $x = L_A$ and the other is the compensating winding and is located between $x = -L_C$ and $x = 0$. The

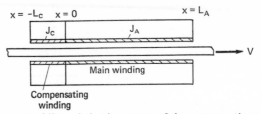

Fig. 57. Compensated linear induction motor of the compensating-winding type.

compensating winding has only two poles and hence its length L_C is much shorter than length L_A of the main winding.

Let current of the main winding be expressed by J_A. Then vector potential in the air gap produced by the main winding within length L_A of the iron core is given by Eq. (180), which is rewritten below.

$$\phi_{3A}(x,y) = \mu_3 J_A \left\{ \frac{G(-k_A,y)}{H(-k_A)} \exp(-jk_A x) \right.$$

$$\left. + \frac{G(\xi_0,y)}{(\xi_0 + k_A)H'(\xi_0)} \exp(j\xi_0 x) \right\}. \tag{189}$$

The compensating winding is located in front of and adjacent to the entry end of the main winding. The vector potential $\phi_{3C}(x,y)$ produced by the compensating winding in the air gap of the main winding is given by Eq. (128), which gives the vector potential in zone III in Fig. 34. The origin of the x coordinate is located at the entry end in Fig. 34, while it is located at the exit end of the compensating winding in Fig. 57. In order to apply Eq. (128) to Fig. 57, x must be replaced by $x + L_C$. Thus, vector potential $\phi_{3C}(x,y)$ produced by the compensating winding in the region $0 < x < L_A$ is given by

$$\phi_{3C}(x,y) = \mu_3 J_C \frac{\exp\{-j(\xi_0 + k_C)L_C\} - 1}{\xi_0 + k_C}$$

$$\times \frac{G(\xi_0,y)}{H'(\xi_0)} \exp\{j\xi_0(x + L_C)\}. \tag{190}$$

This is the exit-end-effect wave of the compensating winding, which is transmitted to the air gap of the main winding. The vector potential in the air gap of the main winding is the sum of $\phi_{3A}(x,y)$ of Eq. (189) and $\phi_{3C}(x,y)$ of Eq. (190). Elimination of the entry-end-effect wave of the main winding, which is given by the second term of Eq. (189), is accomplished if sum of this term and $\phi_{3C}(x,y)$ of Eq. (190) is zero everywhere within length L_A. Thus the condition for compensating the entry-end-effect wave is given by

$$\frac{J_A}{\xi_0+k_A}=\frac{J_C\{\exp(j\xi_0L_C)-\exp(-jk_CL_C)\}}{\xi_0+k_C} \ . \tag{191}$$

Let number of poles of the compensating winding be $P_C=2p_C$. Then $k_CL_C=2p_C\pi$ and, hence, $\exp(-jk_CL_C)=1$. Equation (191) now becomes

$$\frac{J_A}{\xi_0+k_A}=\frac{J_C\{\exp(j\xi_0L_C)-1\}}{\xi_0+k_C} \ . \tag{192}$$

This is the compensation condition which must be satisfied by current J_C of the compensating winding in order to eliminate the end-effect wave in the air gap of the main winding. Then in the air gap of the main winding there remains only the normal wave of the main winding, which is the first term of Eq. (189) and is given by

$$\phi_{3\text{norm}}(x,y)=\mu_3 J_A\frac{G(-k_A,y)}{H(-k_A)}\exp(-jk_Ax) \ . \tag{193}$$

In order to find the smallest possible compensating winding, we must find smallest possible J_C and L_C, which satisfy Eq. (192). The largest value of $|\exp(j\xi_0L_C)-1|$ gives smallest value for J_C. When the motor speed is very high, ξ_0 is almost pure real, and is derived from Eq. (39) as

$$\xi_0\doteqdot-\frac{\pi}{\tau_e}$$

$$\doteqdot-\frac{\pi}{\tau_A(1-s)} \ , \tag{194}$$

where τ_A is the pole pitch of the main winding and slip s refers to the main winding. When ξ_0 is pure real, maximum value of $|\exp(j\xi_0L_C)-1|$ is 2, which is given by

$$\xi_0L_C=n\pi \ , \tag{195}$$

where n is a negative odd integer. Then Eq. (192) becomes

$$J_C=\frac{J_A}{2}\frac{\xi_0+k_C}{\xi_0+k_A} \ . \tag{196}$$

This is now the compensation condition which gives the smallest J_C. On comparison with Eq. (184), J_C is one-half of the compensating current J_B of the compensated linear induction motor of the two-windings type. Length L_C of the compensating winding is much shorter than length L of the compensated linear induction motor of the two-windings type. It can be concluded that the compensated linear induction motor of the compensating winding type is much more economical.

If n in Eq. (195) is chosen as -1, then Eq. (194) gives $L_C=\tau_A(1-s)$.

If number of poles of the compensating winding is 2, then the pole pitch is $\tau_C = (1/2)\tau_A(1-s)$. Similarly, for $n = -3$, we get $L_C = 3\tau_A(1-s)$ and $\tau_C = (3/2)\tau_A(1-s)$. In summary, we have for case $n = -1$

$$L_C = \tau_A(1-s), \qquad \tau_C = \frac{1}{2}\tau_A(1-s), \tag{197}$$

and for case $n = -3$

$$L_C = 3\tau_A(1-s), \qquad \tau_C = \frac{3}{2}\tau_A(1-s). \tag{198}$$

These cases give the two smallest compensating windings because other values of n give a longer L_C; it appears that case $n = -1$ gives the smallest value. However, in selecting one of the two cases, we have to take into consideration current J_C of Eq. (196); the smaller J_C is of course desirable. The following fact also has to be considered. In these equations, s indicates the slip at which the compensation condition of Eqs. (192) or (196) is satisfied, and the slip refers to the main winding. The value of slip s is usually small. Neglecting s in comparison with 1 for case $n = -1$, we get $\tau_C \doteqdot \tau_A/2$. The synchronous speed of the compensating winding is then one-half that of the main winding. In small slip region for the main winding, the compensating winding functions as an induction generator. For case $n = -3$, we have $\tau_C \doteqdot 3\tau_A/2$. The synchronous speed of the compensating winding is then higher than that of the main winding. Hence, in the entire slip range of motor action of the main winding, the compensating winding performs motor action. In choosing one of the two cases the above-mentioned factors must be taken into consideration. It might be possible to say that case $n = -3$ of Eq. (198) is preferred case.

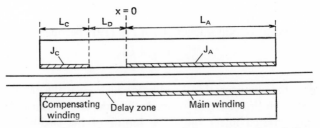

Fig. 58.   Compensated linear induction motor of the compensating-winding type with delay zone.

The compensation condition given by Eqs. (192) or (196) includes both the amplitude condition and the phase condition. The amplitude condition can be met by controlling the number of turns of the compensating winding, since J_C in Eqs. (192) or (196) really indicates ampere turns per meter and not current itself. The phase condition is more difficult to satisfy and an economical method of satisfying it is needed. Figure 58 shows a schematic view of such a method for satisfying the phase condition. In regard to the main winding and the compensating winding, there is no deviation from the explanation of the compensated linear induction motor in Fig. 57, except for the delay zone provided between the compensating winding and the main winding. There are iron cores in the delay zone which are parts of the continuous iron cores but have neither slots nor primary windings. The iron cores function as a wave guide which transmits a travelling wave from the compensating-winding zone to the main winding zone. The exit-end-effect wave generated by the compensating winding is delayed in phase and also slightly attenuated in amplitude while being transmitted through the delay zone. Let length of the delay zone be L_D as shown in Fig. 57. The forward exit-end-effect wave generated by the compensating winding is given by Eq. (190) with x replaced by $x+L_D$. Thus we have

$$\phi_{3C}(x, y) = \mu_3 J_C \frac{\exp\{-j(\xi_0+k_C)L_C-1\}}{\xi_0+k_C}$$
$$\times \frac{G(\xi_0, y)}{H'(\xi_0)} \exp\{j\xi_0(x+L_D+L_C)\} . \tag{199}$$

In order to eliminate the end-effect waves in the main winding zone, the sum of the second term of Eq. (189), which is the entry-end-effect wave of the main winding, and $\phi_{3C}(x, y)$ of Eq. (199), which is the forward exit-end-effect wave from the compensating winding, must be zero everywhere within the main winding zone. Thus, we get the following compensation condition

$$\frac{J_A}{\xi_0+k_A} = \frac{J_C \exp(j\xi_0 L_D)}{\xi_0+k_C}$$
$$\times \{\exp(j\xi_0 L_C)-\exp(-jk_C L_C)\} . \tag{200}$$

When number of poles of the compensating winding is $P_C=2p_C$, $k_C L_C = 2p_C\pi$ and, hence, $\exp(-jk_C L_C)=1$. Then the compensation condition of Eq. (200) becomes

$$\frac{J_A}{\xi_0+k_A} = \frac{J_C \exp(j\xi_0 L_D)\{\exp(j\xi_0 L_C)-1\}}{\xi_0+k_C} . \tag{201}$$

Comparing this equation with Eq. (192), it is noticed that J_C in Eq. (192) is replaced by $J_C \exp(j\xi_0 L_D)$ in Eq. (201). For high-speed motors ξ_0 is given approximately by Eq. (194). The factor $\exp(j\xi_0 L_D)$ then gives phase delaying only and does not give attenuation. Thus by adjusting length L_D of the delay zone, the phase condition of the compensation condition of Eq. (192) can be met without adjusting the phase of current J_C. The delay zone makes it possible to connect both the main and compensating windings to the same power source. They may be connected either in series or in parallel. Which one is preferred, series connection or parallel connection, will be explained in the next section.

13.4. CALCULATIONS OF THE PERFORMANCE OF THE COMPENSATED LINEAR INDUCTION MOTOR

In preceding sections, theories of the compensated linear induction motor, developed in our laboratory, were explained. There are two types of compensated linear induction motors, the two-windings type and the compensating-winding type. As explained in Section 13.2, the two-windings type has two windings of equal length which, therefore, need very large kVA in comparison with the motor output. This type of motor is uneconomical, hence its performance calculation will not be treated here.

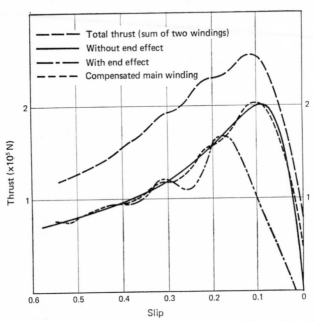

Fig. 59. Thrust-versus-slip characteristics of a compensated linear induction motor (motor D).

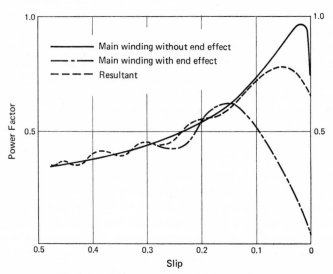

Fig. 60. Power factor-versus-slip characteristics of a compensated linear induction motor.

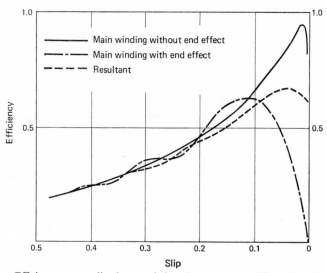

Fig. 61. Efficiency-versus-slip characteristics of a compensated linear induction motor.

The compensated linear induction motor of the compensating-winding type is much more superior, and its performances will be calculated here to show its superiority over the uncompensated linear induction motor.

Motor D in the table in Appendix VI is the motor used in performance calculation. The performance of Motor D under constant voltage drive was calculated according to the method explained in Chapter 11. The results are shown in Figs. 59, 60 and 61; Fig. 59 shows thrust-versus-slip characteristic curves. The solid line indicates the thrust of the normal wave only, and the broken line indicates the thrust of the normal and end-effect waves working together. In the region with a slip of 0–0.2, thrust is reduced considerably by the end effect. Figure 60 shows the power factor-versus-slip characteristic curves. The power factor without the end effect expressed by the solid line can attain very high values. However, the end effect reduces the power factor tremendously in the region with a slip of 0–0.2. The same trend is observed in Fig. 61 showing efficiency-versus-slip characteristic curves, which are also considerably degraded. The degradation of performance due to the end effect is quite large and, hence, it might be difficult to apply this linear induction motor to practical usages. It might be necessary to eliminate performance degradation by the compensation method explained in Section 13.3.

Slip s, where the compensation condition of Eq. (196) is satisfied, is assumed to be 9%. This is a very-high-speed motor and hence Eq. (194) is valid for it. Two poles are chosen for the number of poles of the compensating winding. Pole pitch of the compensating winding is then given by Eq. (198) as $\tau_C = 2.05$ m. In Eq. (196) $k_A = \pi/1.5$, $k_C = \pi/2.05$ and $\xi_0 = -2.3 + j0.0032$. Equation (196) then gives $J_C = (1.86 + j0.02)J_A$. This means that J_C is almost in phase with J_A and the compensating winding may be connected in series with the main winding. Amplitude matching of the two currents is accomplished by adjusting number of turns of each coil. It was assumed that the two windings were connected in parallel, or that they were supplied from separate power sources. The compensation condition of Eq. (196) is satisfied at one slip value only for one compensating winding and, therefore, the selection of this slip is important. The slip must lie in the small slip region which is a normal running region. As slip deviates from this slip, the deviation from the compensation condition begins and the degradation of performance begins to appear. However, when the two windings are connected in parallel, the current in the compensating winding remains rather constant while current in the main winding changes as slip changes. A rather constant current in the compensating winding keeps the deviation from the compensation condition small in the wider slip range above and below the slip of perfect compensation. Performance was calculated for the compensated linear induction motor thus determined, with the two windings connected in parallel. The results are shown in Figs. 59, 60 and 61, and are superposed on the characteristic curves of the uncompensated linear induction motor. In Fig. 59, the broken line marked "compensated main winding" gives the thrust produced by the main winding compensated by the compensating winding. Several times this curve crosses the solid line marked "thrust of main winding without end effect" which indicates the thrust produced by the normal wave of the main winding only. The slip for the last crossover point is 9%, where the compensation condition of Eq. (196) is completely satisfied. Although the compensation condition is not satisfied at any slip other than 9%, thrust deviation from the solid-line curve is relatively small, and the thrust-versus-slip characteristics are improved considerably. For example, at a slip of 5% thrust is increased 4 times by the compensation and positive thrust is still developed at synchronous speed. At synchronous speed the compensation condition is not completely satisfied and, hence, a part of the end-effect wave travelling in the gap remains and produces positive thrust at synchronous speed. The broken line marked "total thrust of main and compensating winding" indicates total thrust produced by the compensated linear induction motor. The total thrust is the sum of the thrust produced by the compensating winding and thrust given by the curve marked "thrust of compensated main winding." The thrust-versus-slip characteristics of the entire motor are very much superior to those of the uncompensated main winding. If we choose a ratio of 2 between the maximum thrust and load thrust, then for the uncompensated motor thrust is 0.86×10^4 N and slip is 9%, while for the compensated motor thrust is 1.26×10^4 N and slip is only 1.2%. The thrust of the compensated motor increases to about 1.5 times and slip decreases to about 0.12 times. In Fig. 60 the broken line marked "resultant" indicates the resultant power factor of the compensated linear induction motor. The improvement of the power factor over that of the main winding with the end effect is considerable in small slip region. In Fig. 61 the broken line marked "resultant" indicates the overall efficiency of the compensated linear induction motor. Efficiency is improved in the small slip region over that of the uncompensated motor indicated by the broken line marked "main winding with end effect."

As an experimental example, motor E in the table of Appendix VI was provided the compensating winding, and compared with the uncompensated motor, the motor output increased by more than 80%, while the total kVA increased only by 40%.

The effectiveness and usefulness of the compensation is most distinctively demonstrated by increased thrust per pole of the compensated linear induction motor. Figure 62 gives thrust per pole-versus-slip curves for motor E. This motor has 6 poles and the curve marked "6 poles" gives thrust per pole of the uncompensated motor E, the curve marked "8 poles" gives thrust per pole of the uncompensated 8-pole motor, and the curve marked "12 poles" gives thrust per pole of the uncompensated 12-pole motor. It can be seen that thrust per pole is increased with the addition of poles of the ordinary winding. The curve marked "compensated" gives the thrust per pole of the compensated motor E with the addition of two-pole compensating winding. The curve comes close to the curve marked "without end effect," which gives thrust per pole of motor E not under the influence of the end effect. It should be noticed that the maximum thrust per pole is increased by compensation to about 2.4 times that of the uncompensated motor E with 6 poles.

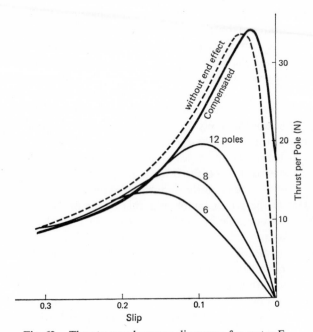

Fig. 62. Thrust per pole-versus-slip curves for motor E.

Chapter 14

A Linear Induction Motor of the Wound-secondary Type*

14.1. INTRODUCTION

So far we have concentrated on linear induction motors whose secondary is a simple homogeneous metallic sheet. A number of researchers have advocated composite secondaries, which are secondary sheets made up of several kinds of metals, one of which is usually iron. A sandwiched secondary, for example, consists of an iron plate covered with copper or aluminum surface sheets, as shown in Fig. 63. Figure 64 shows another kind of composite secondary which consists of aluminum plate and steel slugs imbedded in slots cut in the aluminum plate. The main object of the composite secondary is to decrease the reluctance of the magnetic path and the exciting current. However, as shown in Section 10.3, the unlaminated

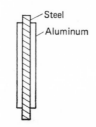

Fig. 63. Sandwiched-type composite secondary.

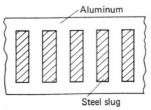

Fig. 64. Composite secondary made of an aluminum plate with steel slugs.

* Reported originally in refs. 19 and 27.

iron plate presents a skin effect to a large extent even for slip frequency and, hence, motor performance is not improved, but, rather, degraded. On the other hand, as was shown in Chapter 7, linear induction motors with a nonferrous sheet can have a very high power factor and very high efficiency in spite of a larger air gap when motor speed is high and the end effect is eliminated. Composite secondaries also have other drawbacks, such as complicated structures, higher cost and lateral magnetic pull.

Furthermore, it seems rather strange that a wound secondary has never been proposed for linear induction motors. The wound-rotor-type is one of the commonest types of rotary induction motors and makes speed control and starting control of induction motors possible using secondary resistance control and other means. The superior controlability of the wound rotor type induction motor more than compensates for slight disadvantage originating from the more complicated structure of the rotor, and this trend holds also for the linear induction motor of the wound-secondary type.

The biggest advantage gained with the wound secondary is complete elimination of the end effect, which occurs in the linear induction motor with the sheet secondary. This will be explained later in this chapter.

14.2. STRUCTURE OF A LINEAR INDUCTION MOTORS OF THE WOUND-SECONDARY TYPE

The primary side of the linear induction motor of the wound-secondary type is the same as that of the sheet-secondary type. The secondary conductors, are not simple conductive sheets of aluminum or copper, but are wound into an insulated winding of poly-phases and it is not necessary for the secondary to be composite. The composite secondary means here that a combination of ferrous and nonferrous metals is used in structure of the secondary. The purpose of the composite secondary is to allow the iron parts to provide low reluctance to the magnetic flux and the nonferrous metals to provide low resistance to the secondary electric current. However, as explained in Chapter 7, high-speed linear induction motors can have a high power factor and high efficiency in spite of a longer air gap, if the end effect is eliminated. It should be emphasized here once more that the high-speed linear induction motor is inherently a high power factor motor and the low-speed linear induction motor is inherently a low power factor motor. It is the end effect that makes the power factor of the high-speed motor low. It seems that this situation is generally misunderstood and it is thought that linear induction motors are generally low power factor motors. As will be explained later in this section, the wound secondary eliminates the end effect completely. The composite secondary then becomes unnecessary for the improvement of the power factor and involves too many complications in the structure of the secondary.

The wound secondary, which we proposed, consists of nonferrous, individually insulated metallic conductors with no iron parts. Their structures are shown in Fig. 65. Each conductor has a rectangular cross section and is covered with thin insulation. The conductors are densely arranged in a plane and are connected by a polyphase winding. Figure 65 (a) shows a general view, (b) cross section of the two-layer winding and (c) cross section of the one-layer winding. Since the cross section of conductors is thick, the wound secondary can be as strong as the continuous sheet secondary. There is no problem concerning mechanical strength of the wound secondary. A more elaborate wound-secondary structure may increase the cost of the linear induction motor beyond that of the rotary induction motor. The higher cost might be compensated for by the advantages gained by using the wound secondary. These advantages include speed control and starting control using the secondary resistance control and other means. As will be explained later, the wound secondary eliminates the end effect which occurs in linear induction motors with a sheet secondary. This is an additional major advantage of the wound-secondary-type linear induction motor. It should be mentioned here that the composite wound secondary need not be and should not be used. As

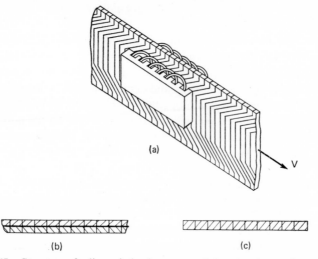

Fig. 65. Structure of a linear induction motor of the wound-secondary type.

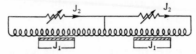

Fig. 66. Schema of a linear induction motor of the wound-secondary type (single-phase representation).

Fig. 67. Schema of a linear induction motor of the wound-secondary type in a case of partial coverage.

was stated above, the existence of unlaminated iron on the secondary side does not improve the performance but rather degrades it and makes the structure of the secondary much more complicated. The wound secondary should also be nonferrous. The simplest structures of a nonferrous wound secondary will be proposed later in this section. This type of wound secondary is the most suitable for a linear induction motor whose primaries are installed on-ground and whose secondaries are installed on-vehicles. In secondary-on-vehicle, primary-on-ground-type trains, speed control and starting can be performed by adjusting the secondary resistance installed on vehicles under constant frequency drive. Even contactless current collection is possible through the wound secondary. Power consumption on the vehicles is then met by the secondary power which would otherwise be discarded as a heat loss.

Although the wound secondary eliminates the kind of end effect that has been treated all through the preceding chapters, another kind of transient phenomenon occurs which does some harm in this type of linear induction motor. There are merits and demerits as well as new problems in this type of linear induction motor which will be explained and analysed in this chapter. It should be pointed out that the merits by far outweigh the demerits and more attention should be paid to the wound-secondary type of linear induction motor.

14.3. End Effect and Transient Phenomenon of Linear Induction Motors of the Wound-secondary Type

Figure 66 shows a schematic drawing of the wound-secondary-type motor applied to train propulsion. The wound secondary is represented by a single phase circuit, and the primary is represented by one side, although there may be actually two primary sides. Since the primary current J_1 and the second current J_2 are uniformly distributed over the entire length of the primary core, the magnetic field must be also uniformly distributed in the air gap. When the secondary is a continuous sheet, secondary current distribution can be nonuniform; the secondary current density near the entry end may be much higher than that near the exit end. This is the cause of the end effect of the linear induction motor of the continuous-sheet-secondary type. When the secondary is the wound type and conductors of the same phase are connected in series, the secondary current is sinusoidally distributed in space and then it is impossible for the end effect to occur. There remains then only the normal travelling wave in the air gap. All the phenomena discussed in the preceding chapters in connection with the end effect would disappear entirely, and all the normal performances of the induction motor would be restored, including a higher power factor in case of the high-speed linear induction motor. The wound secondary is a perfect and comprehensive solution to all the problems and difficulties associated with the end effect.

The end effect is a stationary phenomenon viewed from the x coordinate axis fixed to the primary side. All expressions of the end-effect phenomena in the preceding chapters contain functions of time in the form of $\exp(j\omega t)$, and not in the form of $\exp(-\alpha t)$. This means that the end effect is a stationary phenomenon. When the end effect is viewed from the coordinate axis fixed to the secondary, it is a transient phenomenon. End-effect current in the secondary decays with time and expressions of the end effect phenomena contain a time function in the form of $\exp(-\alpha t)$. However, if the entire length of the primary core is covered by a single section of the wound secondary, both primary and secondary currents are sinusoidally distributed over the entire length of the primary core. Then the magnetic field is also sinusoidally distributed over the entire length of the primary core, and no end effect exists.

In the wound-secondary-type motor transient phenomenon of a different kind appears. As shown in Fig. 67, when the primary core is partially covered by the secondary, it causes a transient phenomenon. In a portion of the iron core, which is covered by the secondary for the first time, the air-gap field is weakened. Even after the entire length of the iron core is covered, it takes some time for the air-gap field to reach a steady field pattern. This phenomenon is caused by the entry end of the secondary, while the end effect is caused by the entry end of the primary. This phenomenon is a transient with respect to time, irrespective of the viewing side, and it occurs only when the end of the secondary reaches the iron core for the first time.* This transient phenomenon is caused by the entry end of the wound secondary and also by the junction of the adjacent sections of the wound secondary as shown in Fig. 67(b). Let the transient be called "wound-secondary-end transient." The transient phenomenon may have a rather large time constant and may affect the performance of the wound-secondary-type motor.

14.4. ANALYSIS OF WOUND-SECONDARY-END TRANSIENT

In the analysis of preceding chapters, the primary winding is represented by the equivalent current sheet. Similarly, the secondary winding of the wound secondary can be represented by its equivalent current sheet. The field equations for the air gap of the wound-secondary linear induction motor are then the same as those given in Chapter 3 and are rewritten below.

$$\frac{\partial b}{\partial x} = \frac{\mu_0}{g}(j_1 + j_2),$$ (202)

$$\frac{\partial e_2}{\partial x} = \frac{\partial b}{\partial t} + v\frac{\partial b}{\partial x}.$$ (203)

They are the same as Eqs. (8) and (9), respectively. In case of the constant current drive, the primary current j_1 is assumed to be expressed by

$$j_1 = J_1 \exp\left\{j\left(\omega t - \frac{\pi}{\tau}x\right)\right\}.$$ (204)

J_1 is maximum current density of the equivalent primary current sheet and is related to the actual primary current I_1, of the three-phase winding by Eq. (6); rewriting we have

$$J_1 = \frac{3\sqrt{2}\,w_1k_{w1}I_1}{p_1\tau} \quad \text{(A/m)}.$$ (205)

Similarly, the actual secondary current i_2 (instantaneous value) is represented by j_2 of the equivalent secondary current sheet in the theory and they are related approximately by

$$j_2 = \frac{m_2w_2}{p_2\tau}i_2 \quad \text{(A/m)},$$ (206)

where m_2 is number of phases, w_2 number of turns in series per phase, p_2

---

* Another kind of transient phenomenon occurs when the secondary end is leaving the iron core, but its influence is not so serious, and, hence, this phenomenon is not discussed here.

number of pole pairs and τ_2 pole pitch for the wound secondary. Equation (206) is a better approximation when number of phases is larger. e_2 in Eq. (203) is the induced electromotive force per unit length of a secondary conductor. When secondary leakage reactance is much smaller than the secondary resistance and is neglegible, then following equations holds

$$e_2 = \frac{R_2 p_2}{2 w_2 D p_1} i_2 = \rho j_2 , \qquad (207)$$

$$\rho = \frac{R_2 p_2{}^2 \tau}{2 m_2 w_2{}^2 p_1 D} , \qquad (208)$$

where R_2 is secondary resistance per phase, D the iron stack width and p_1 the number of primary pole pairs. It should be noticed that pole pitch τ is the same for both primary and secondary and p_2 is larger than p_1, as can be seen in Figs. 66 or 67. From Eqs. (202), (203) and (206) we get

$$\frac{g}{\mu_0} \frac{\partial^2 b}{\partial x^2} - \frac{v_2}{\rho} \frac{\partial b}{\partial x} - \frac{1}{\rho} \frac{\partial b}{\partial t} = \frac{\partial j_1}{\partial x} . \qquad (209)$$

When j_1 of Eq. (204) is the sole exciting source, the steady state solution of Eq. (209) can be expressed by the following form:

$$b_{\mathrm{S}} = B_{\mathrm{S}} \exp\left\{ j\left(\omega t - \frac{\pi}{\tau} x \right) \right\} . \qquad (210)$$

Substituting Eq. (210) into Eq. (209), we get

$$B_{\mathrm{S}} = I_{\mathrm{S}} \Big/ \left(\frac{\tau \omega - \tau v_2}{\rho \pi} - j \frac{g \pi}{\mu_0 \tau} \right) . \qquad (211)$$

The general solution of Eq. (209) was already obtained as Eq. (26), λ_n, kn_1 and kn_2 in Eq. (26) must be chosen properly so that conditions imposed by the linear induction motor of the wound secondary are satisfied. Since spacial distributions of both the primary current j_1 and secondary current j_2 are sinusoidal, spacial distribution of the flux density in the air gap must be also sinusoidal and it may be assumed that it consists of space harmonics of wavelength $2\tau/n$, n being arbitrary integers. Then the general solution of Eq. (26) may now be written as follows:

$$b(x,t) = B_{\mathrm{S}} \exp\left\{ j\left(\omega t - \frac{\pi}{\tau} x \right) \right\}$$
$$+ \sum_n B_n \exp(\lambda_n t) \exp\left(-j \frac{n\pi}{\tau} x \right) . \qquad (212)$$

Substituting Eq. (212) into Eq. (209), we have

$$\lambda_n = -\frac{n^2\pi^2\rho g}{\mu_0\tau^2} + j\frac{n\pi v}{\tau} . \tag{213}$$

Substituting Eq. (213) into Eq. (212), the general solution takes now the following forms:

$$b(x,t) = B_s \exp\left\{ j\left(\omega t - \frac{\pi}{\tau}x\right)\right\}$$

$$+ \sum_n B_n \exp\left(-\frac{n^2\pi^2\rho g}{\mu_0\tau^2}t\right)\exp\left\{jn\left(\frac{\pi v}{\tau}t - \frac{\pi}{\tau}x\right)\right\} . \tag{214}$$

The first term is the normal wave and the second term is the transient components. The transient components are all travelling waves and their speed is v (motor speed) irrespective of the order n, and they decay with respect to time. The time constant, which controls the decay, is given by

$$T_n = \frac{\mu_0\tau^2}{n^2\pi^2\rho g} . \tag{215}$$

The transient components can be determined from the field distribution $b(x,0)$ at time$=0$. In high-speed motors τ is larger and time constant T_n is larger and the transient components decay rather slowly. When the leading end of the wound secondary plunges into the air gap, the magnetic field in the air gap is weakened and it takes some time for the magnetic field to build up. When the motor speed is high, it is reasonable to assume that the magnetic field is not yet recovered at all even at the moment when the entire length of the iron core is covered by the secondary. Then as the initial condition $b(x,0)$, we may adopt $b(x,0)=0$ everywhere. Equation (214) gives then the following:

$$B_s \exp\left(-j\frac{\pi}{\tau}x\right) + \sum B_n \exp\left(-j\frac{n\pi}{\tau}x\right) = 0, \tag{216}$$

from which we get

$$B_n = -B_s \quad \text{for} \quad n=1, \qquad B_n=0 \quad \text{for} \quad n\neq 1 . \tag{217}$$

And now Eq. (214) becomes

$$b(x,t) = B_s \exp\left\{ j\left(\omega t - \frac{\pi}{\tau}x\right)\right\}$$

$$- B_s \exp\left(-\frac{t}{T_1}\right)\exp\left\{j\left(\frac{\pi v_2}{\tau} - \frac{\pi}{\tau}x\right)\right\} , \tag{218}$$

where

$$T_1 = \frac{\mu_0 \tau^2}{\pi^2 \rho g}.\tag{219}$$

For $\tau = 1.5$ m, $g = 0.04$ and $\rho = 10^{-5}\Omega$, $T_1 = 0.72$ s. When motor speed v is 100 m/s, it takes $5/100 = 0.05$ s for the entire length of the iron core to be covered by the wound secondary speeding at 100 m/s. Since $0.05 \ll T_1 = 0.75$, the above-mentioned assumption of $b(x,t) = 0$ is reasonable for this case.

The time constant T_1 of Eq. (219) is comparatively large, and it takes some time for the motor to begin its thrust-producing function after the wound secondary has entered its air gap. $3T_1$ would be a good measure of the necessary time, and then for the above-mentioned example the vehicle would have run $3 \times 0.72 \times 100 = 216$ m before the linear induction motor begins its normal thrust-producing function. This means considerable loss of thrust and output. It is then necessary to eliminate the wound-secondary-end transient.

Figure 68 shows a schematic drawing of the method of eliminating the wound-secondary-end transient. In this figure the wound secondary is also drawn in a single-phase representation. There are a series of primaries and a series of secondaries. The secondaries are arranged in a row without intervals, while the primaries are installed at equal distances. When the length of one section of the secondary is equal to one installation pitch of the primaries or integral of it, the portion of a section of the secondary which is facing a primary core or cores is always constant, irrespective of the relative position between the primary and the secondary, as shown in Fig. 68. Then total flux linkage of one section of the secondary is always constant and hence the wound-secondary-end transient cannot occur.

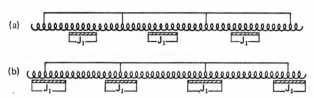

Fig. 68. Schema of a linear induction motor of the wound-secondary type for eliminating the wound-secondary-end transient.

14.5. Starting, Speed Control and Contactless Power Collection of High-speed Trains by Means of Linear Induction Motors of the Wound-secondary Type

Having eliminated the end effect and the wound-secondary-end transient, normal operating performance can be expected from the linear induction motor of the wound-secondary type. It is especially suitable for high-speed train propulsion. At higher speeds a high power factor and high efficiency can be expected from the linear induction motor, in spite of a large air gap, as was explained in Chapter 7.

An essential problem concerning the application of linear induction motors to train propulsion is which type of linear induction motor is most advantageous, the primary-on-ground, secondary-on-vehicle type or the primary-on-vehicle, secondary-on-ground type. Each type has its merits and demerits, and both types are under serious investigation in super-high-speed train projects in many countries. This book is not the appropriate place for discussing this problem, however, it should be pointed out that the linear induction motor of the wound-secondary type is the most suitable to the primary-on-ground, secondary-on-vehicle type of applications. The wound-secondary-on-vehicle enables speed and starting control by means of secondary resistance adjustment. This is the lowest priced speed and starting control system of the induction motor, and it eliminates the need for expensive variable frequency power supply.

A great merit of the primary-on-ground, secondary-on-vehicle type is that it makes power collection on the side of vehicles unnecessary, power being supplied directly to the primaries installed on ground. However, power consumption on vehicle still must be covered by some means. In induction motors, power is transmitted through the air gap to the secondary side. Concerning the secondary power, a portion is converted into secondary copper loss, and thus the following relations hold.

Mechanical power $=(1-s)W_2$,

Secondary copper loss $=sW_2$.

Here s is the slip and W_2 is the secondary power. sW_2 portion need not be converted into heat, but may be utilized to meet power demand on vehicles. When the propulsion power of a high-speed train is 3000 kW per car at a slip of 4%, then the power which can be consumed in the secondary

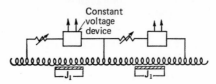

Fig. 69.   Schema of a linear induction motor of the wound-secondary type for supplying power to loads-on-vehicle.

can be a little more than 120 kW. This scheme of covering the power demand on vehicles improves overall power efficiency. The greatest merit of the scheme is that it makes contactless power collection possible. Current or power collection is one of the most difficult problems associated with high-speed electric trains. A combination of pantographs and contact wire may not be used for very fast trains, so that contactless power collection is most desirable. There lies the possibility of solving the most difficult problem in this type of linear induction motor, which might accord an excellent propulsion system at the same time. Figure 69 is a skeleton diagram of such a scheme, where the linear induction motor of the wound-secondary type provides propulsion and power supply to the vehicle at the same. In the secondary circuit there is a speed control device, such as adjustable resistance, and a constant voltage device, which constitutes a constant voltage power supply for power consumption on vehicles.

14.6. EXAMPLE OF A LINEAR INDUCTION MOTOR OF THE WOUND-SECONDARY TYPE

The results of performance calculations for a linear induction motor of the wound-secondary type will be shown. The dimensions of the primary iron core is 3.9 m in length and 0.3 m in width. The secondary is 30-m long. The 30-m length corresponds to the length of one coach of a train. The dimension of a cross section of a secondary conductor shown in Fig. 65(b) is 8 mm in thickness and 30 mm in width. So the thickness of the secondary winding is about 17 mm. Total length of the air gap including the secondary winding is 60 mm. A series of primaries and secondaries are installed, as shown in Fig. 68, in order to eliminate the transient phenomenon explained in the preceding section. Performance for one pair consisting of a primary and a secondary were calculated, and the result is shown in Fig. 70. It can be seen from the figure that both the maximum values of efficiency and power factor are high, in spite of the large gap length of 60 mm, although the motor characteristics have a trend of higher secondary resistance. It can be concluded that the linear induction motor of noncomposite, wound-secondary type shows good performance which is suitable for the propulsion of a high-speed trains as proposed in the preceding stecion of this chapter.

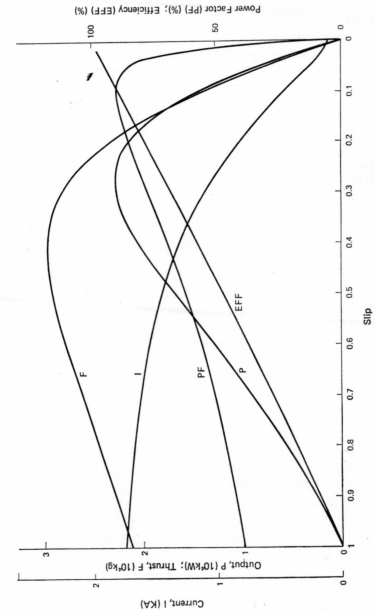

Fig. 70. Characteristic curves of a linear induction motor of the noncomposite wound-secondary type.

Appendix I

Determination of Secondary Resistance

In analyses in preceding chapters the secondary resistance is designated by ρ_S and the secondary conductance by σ_S. They are surface resistivity or surface conductivity, respectively, and are given by $\rho_S =$ (volume resistivity) /(sheet thickness) $= \rho/2b$ (Ω) and $\sigma_S =$ (volume conductivity) $\times$ (sheet thickness $= 2b\sigma$ (℧), ρ_S is resistance of the secondary sheet 1-m long and 1-m wide.

It is assumed then that the secondary current has only the z zomponent in the air-gap zone, as shown in Fig. A-1(a). In effect, this means that the secondary sheet has resistance only in the air-gap zone and its end portion or overhang has no resistance. Actually the sheet ends are extensions of the secondary sheet in the air-gap zone and are made of the same metal and have resistance. The surface resistivity ρ_S of the secondary sheet must include the resistance of the sheet ends also. The proper inclusion of sheet end resistance is a rather complicated problem.

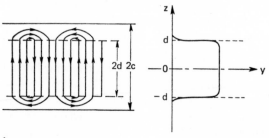

(a) Secondary current pattern.
    $2d =$ thickness of iron stach;
    $2c =$ width of secondary sheet.

(b) transverse distribution of
    magnetic flux density.

Fig. A-1. Secondary current pattern and transverse distribution of magnetic flux density.

Another problem arises in regard to this. The secondary current flow is not totally directed in the z direction in the air-gap zone but has also an x component. A portion of the secondary current closes it path within the air-gap zone as shown in Fig. A-2(a). A corresponding distribution of the magnetic field in the z direction is shown in Figs. A-1(b) and A-2(b). The uniform distribution in Fig. A-1(b) corresponds to the parallel current flow within the air gap zone, while nonuniform flux distribution in Fig. A-2(b) corresponds to nonparallel current flow. The nonuniform distribution of the secondary current and magnetic flux in the z direction is sometimes called the transverse end effect.

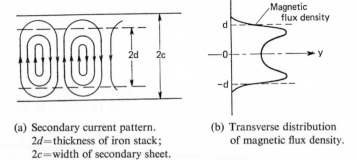

(a) Secondary current pattern.
    $2d$=thickness of iron stack;
    $2c$=width of secondary sheet.

(b) Transverse distribution
    of magnetic flux density.

Fig. A-2. Secondary current pattern and transverse distribution of magnetic flux density (actual case).

Although it is sometimes called the end effect, the transverse end effect is quite different in its origin and nature from the end effect, which is the main subject of the preceding chapters. It is caused neither by the relative motion between the primary and secondary nor does it involve transient phenomenon of any kind. It is simply a steady state phenomenon concerning nonuniform distribution of current and magnetic flux in the transverse direction. We shall call it transverse nonuniformity.

When an attempt is made to determine the effective resistance of the secondary sheet, it is necessary to take into account the transverse nonuniformity. The transverse nonuniformity is a factor, which has some influence on the effective secondary resistance and also on motor performance.

Strictly speaking, the end effect and the transverse nonuniformity influence each other, and rigorous solutions could only be obtained by three-dimensional analysis. However, such analysis would be very complicated. As a good approximation, they might be treated separately and results could be superposed.

The transverse nonuniformity was investigated by several research-ers who obtained analytical solutions concerning it and indicated how it influences motor performances. Here their results will be utilized and somewhat extended in order to derive analytical expressions of effective secondary resistance and effective secondary current. Most of the following analysis will follow the argument in the paper by H. Bolton.[17]

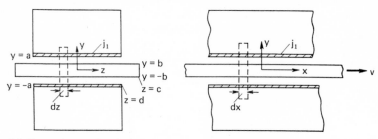

(a) Transverse sectional view.   (b) Longitudinal sectional view.

Fig. A-3. Sectional view of a linear induction motor.

Figure A-3 is the model of the linear induction motor on which the following analysis is based. Figure A-3(a) gives the transverse sectional view and (b) gives the longitudinal view. The coodinate axes are indicated in the figures and are fixed to the secondary. Applying Maxwell's equations to the small loops in the figures, the following field equation is derived:

$$\frac{\partial^2 b}{\partial x^2} + \frac{\partial^2 b}{\partial z^2} - \frac{\mu_0}{2a\rho_s}\frac{\partial b}{\partial t} = -\frac{\mu_0}{2a}\frac{\partial(2j_1)}{\partial x}, \tag{A-1}$$

where

ρ_s = surface resistivity of the secondary sheet

= (volume resistivity)/(thickness of the sheet)

$$= \frac{\rho}{2b} = \frac{1}{2\sigma b} = \frac{1}{\sigma_s},$$

$2a$ = gap length,

$2c$ = secondary sheet width, and

$2d$ = thickness of primary iron stack.

Equation (A-1) then becomes

$$\frac{\partial^2 b}{\partial x^2} + \frac{\partial^2 b}{\partial z^2} - \mu_0\sigma\frac{b}{a}\frac{\partial b}{\partial t} = -\frac{\mu_0}{a}\frac{\partial j_1}{\partial x}. \tag{A-2}$$

Since only steady state is under investigation, all quantities are sinusoidal in x and t. Let b and j_1 be given by

$$b = B \exp\{j(s\omega t - kx)\} ,$$
$$j_1 = J_1 \exp\{j(s\omega t - kx)\} . \qquad (A\text{-}3)$$

Substituting these into Eq. (A-2), we get

$$\frac{\partial^2 B}{\partial z^2} - \left(k^2 + js\omega\mu_0\sigma \frac{b}{a} \right) B = \frac{j\mu_0 k}{a} J_1 . \qquad (A\text{-}4)$$

When there is no transverse nonuniformity, $\partial^2 B/\partial z^2 = 0$ and the solution of Eq. (A-4) is given by

$$B = -\frac{j\mu_0 J_1}{ak} \frac{1}{1+jsG} = -\frac{j\mu_0 J_1}{ak} \gamma^2 , \qquad (A\text{-}5)$$

$$\left.\begin{array}{l} \gamma^2 = \dfrac{1}{1+jsG} , \\[2mm] G = \dfrac{2\tau^2 \mu_0 f\sigma_s b}{\pi a} , \\[2mm] k = \dfrac{\pi}{\tau} . \end{array}\right\} \qquad (A\text{-}6)$$

H. Bolton[17] gave the solution of Eq. (A-1) as follows:

$$B = -j\frac{\mu_0}{ak} J_1 \gamma^2 \left\{ 1 + \frac{1-\gamma^2}{\gamma^2} \lambda \left(\frac{\cosh \alpha z}{\cosh \alpha d} \right) \right\} , \qquad (A\text{-}7)$$

$$\lambda = \frac{1}{1+(1/\gamma)\tanh \alpha d \tanh k(c-d)} , \qquad (A\text{-}8)$$

$$\alpha^2 = k^2 + \frac{js\omega\mu_0}{2a\rho_s}$$
$$= k^2(1+jsG) . \qquad (A\text{-}9)$$

B of Eq. (A-7) gives the flux distribution, which takes the transverse nonuniformity into consideration.

It should be pointed out here that the length of the overhang $c-d$ of the secondary sheet is contained only in the factor $\tanh k(c-d)$ in the numerator of Eq. (A-8). This factor approaches 1 asymptotically and increases very slowly after $k(c-d)$ becomes larger than 1.3, for which we have

$$\frac{c-d}{\tau} = \frac{1.3}{\pi} \doteqdot 0.41.$$

It is useless to make the overhang length longer than about 40% of the pole pich τ, for the purpose of reducing ρ_s.

The mean value of flux density B of Eq. (A-7) over the iron stack thickness is given by

$$B_{mean} = \frac{1}{2d} \int_{-d}^{d} B dz$$

$$= -j\frac{\mu_0 J_1}{ak}\gamma^2\left(1+\frac{1-\gamma^2}{\gamma^2}\frac{\lambda}{\alpha d}\tanh \alpha d\right). \qquad (A-10)$$

Substituting the following abbreviation into Eq. (A-10):

$$\frac{\lambda}{\alpha d}\tanh \alpha d = u + jv \qquad (A-11)$$

and rewriting it, we get

$$B_{mean} = -j\frac{\mu_0 J_1}{ak}\frac{1+jsG(u+jv)}{1+jsG}$$

$$= -j\frac{\mu_0 J_1'}{ak}\frac{1}{1+jsG'}, \qquad (A-12)$$

where

$$J_1' = J_1\frac{(1-sGv)^2+s^2G^2u^2}{1-sGv+s^2G^2u}, \qquad (A-13)$$

$$G' = G\frac{1-u-sGv}{1-sGv+s^2G^2u}. \qquad (A-14)$$

Comparison of Eq. (A-5) and (A-12) indicates that the transverse non-uniformity replaces J_1 with J_1' of Eq. (A-13) and G with G' of Eq. (A-14). From Eq. (A-6) it can be seen that the replacement of G with G' is equivalent to the following replacements.

$$\sigma_s' = \sigma_s\frac{1-u-sGv}{1-sGv+s^2G^2u}, \qquad (A-15)$$

$$\rho_s' = \rho_s\frac{1-sGv+s^2G^2u}{1-u-sGv}. \qquad (A-16)$$

These are surface conductivity and surface resistivity, respectively, which take into account both the transverse nonuniformity and resistance of the secondary sheet overhang. The primary current must also be replaced with J_1' of Eq. (A-13), in order to take them into account.

Figure A-4 shows calculated results of J_1'/J_1 and σ_s'/σ_s for the following case:

air gap $2a = 15$ mm,

sheet materials: copper or aluminum,

thickness of sheet: 5 mm,

$d = 45$ mm,

$c = 75$ mm.

It can be seen from the figure that J_1'/J_1 drops very little from 1 and σ'/σ drops considerably from 1. For this example, the correction of J_1 is not necessary and the replacement of σ_s with $\sigma_s' = 0.65\sigma_s$ is a good approximation for slip range $0 \sim 1$.

In all the calculations of the preceding chapters surface conductivity of the secondary sheet was determined by the method derived here, and no correction was made for the primary current J_1.

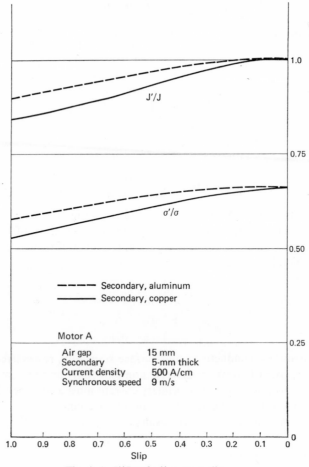

Fig. A-4. J'/J and σ'/σ versus slip.

Calculation of Linear Induction Motor Performance by Means of the Relaxation Method

The relaxation method is a very effective method for the numerical analysis of differential equations. It is very effective in solving Laplace's equation and Poisson's equation. However, it becomes rather cumbersome when it is applied to Maxwell's equation. Y. Ishikawa, working under my guidance, succeeded in solving Maxwell's equations for the air-gap field of the linear induction motor by means of the relaxation method, and reported the results in his master's thesis.[26] Numerical calculation by means of the relaxation method is based on the most rigorous assumption among the three approaches to the problem of the linear induction motor, which were attempted in our laboratory. They are the one-dimensional analysis, the two-dimensional analysis and the numerical calculation by means of the relaxation method. Some of the results obtained by the relaxation method are shown in Figs. A-5 and A-6. The relaxation method gives a two-dimensional distribution of the magnetic field of the air gap in the x-y plane. Figure A-5 shows magnetic flux density distribution on the surface of the primary iron core for motor A.

The influence of the entry-end effect and the exit-end effect are clearly recognized; that is, near the entry end the field is weakened considerably by the entry-end wave, and near the exit end there are dips and peaks due to the exit-end-effect wave. Figure A-6 shows thrust-versus-slip curves for motor A. Thrust-versus-slip curves calculated with the two-dimensional solution by means of Fourier transform shown in Chapter 11 are also drawn for comparison and both results are in good agreement. It should be pointed out that motor A is a low-speed motor and produces thrust even at synchronous speed due to the end effect.

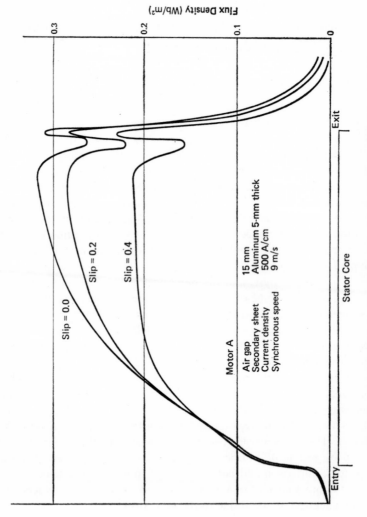

Fig. A-5. Magnetic flux density distribution calculated by the relaxation method.

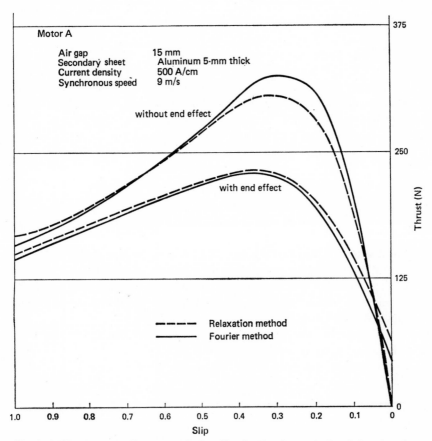

Fig. A-6. Thrust-versus-slip curves calculated by the Fourier method and the relaxation method.

Boundary Conditions

In Chapter 3 field equation solutions were derived, based on the one-dimensional model shown in Fig. 8. The solution of magnetic flux density distribution is given by Eq. (35), which is rewritten below.

$$b = B_s \exp\left\{j\left(\omega t - \frac{\pi}{\tau}x + \delta_s\right)\right\}$$

$$+ B_1 \exp\left(-\frac{x}{\alpha_1}\right) \exp\left\{j\left(\omega t - \frac{\pi}{\tau_e}x + \delta_1\right)\right\}$$

$$+ B_2 \exp\left(\frac{x_1}{\alpha}\right) \exp\left\{j\left(\omega t - \frac{\pi}{\tau_e}x + \delta_2\right)\right\} . \qquad \text{(A-17)}$$

The first term is the normal wave, which is given by Eqs. (15) and (16). The second term is the entry-end-effect wave and the third term is the exit-end-effect wave. They contain the arbitrary parts $B_1 \exp(j\delta_1)$, and $B_2 \exp(j\delta_2)$, respectively, which are to be determined from the two boundary conditions. We shall now find the boundary conditions.

If magnetic flux does not leave the iron core through the back and sides and if there is no fringing of flux at both ends, then flux exits and enters the iron core only within the air gap. Then we have

$$\int_0^L b \, dx = 0 , \qquad \text{(A-18)}$$

where L is the core length. Equation (A-18) may be one of the boundary conditions and it holds in many cases. We shall now seek other boundary conditions.

From Eqs. (9) and (10) the following equation is derived:

$$\rho_s \frac{\partial j_z}{\partial x} = \frac{\partial b}{\partial t} + v \frac{\partial b}{\partial x} . \qquad \text{(A-19)}$$

At the entry end of the air gap the magnetic field is built up very steeply and $|\partial b/\partial x|$ is much larger than $|\partial b/\partial t|$. Except for very low speed,

$$\left| v \frac{\partial b}{\partial x} \right| \gg \left| \frac{\partial b}{\partial t} \right|$$

and Eq. (A-19) becomes

$$\rho_S \frac{\partial j_z}{\partial x} = v \frac{\partial b}{\partial x} . \tag{A-20}$$

If there is no fringing of magnetic flux, it may be assumed that $j_2 = 0$ and $b = 0$ for $x < 0$. Then Eq. (20) gives

$$b \bigg|_{x=0} = \frac{\rho_S}{v} j_2 \bigg|_{x=0} . \tag{A-21}$$

At higher speed the magnetic field is weakened at the entry end and hence the exciting current component of j_1 can be neglected at $x = +0$. Then we have $j_1 = -j_2$ at $x = +0$ and Eq. (A-21) now becomes

$$b \bigg|_{x=0} = -\frac{\rho_S}{v} j_1 \bigg|_{x=0} . \tag{A-22}$$

Equation (A-22) can be one of the boundary conditions.
    From Eq. (8) we get

$$j_2 = \frac{g}{\mu_0} \frac{\partial b}{\partial x} - j_1 . \tag{A-23}$$

At the entry end the term B_2 can be neglected in Eq. (A-17) and then we get

$$\frac{\partial b}{\partial x}\bigg|_{x=0} = -\left\{ j \frac{\pi}{\tau} B_S \exp(j\delta_S) \right.$$

$$\left. + \left(\frac{1}{\alpha} + j \frac{\pi}{\tau_e} \right) B_1 \exp(j\delta_1) \right\} \exp(j\omega t) . \tag{A-24}$$

From Eqs. (A-21), (A-23) and (A-24) we get

$$B_1 \exp(j\delta_1) = \frac{-\frac{\rho_S}{v} \left\{ J_1 + j \frac{g}{\mu_0} \frac{\pi}{\tau} B_S \exp(j\delta_S) \right\} - B_S \exp(j\delta_S)}{1 + \frac{\rho_S}{v} \frac{g}{\mu_0} \left(\frac{1}{\alpha_1} + j \frac{\pi}{\tau_e} \right)} . \tag{A-25}$$

This is the entry-end-effect wave at $x = 0$. It should be noticed that, when v is very large, Eq. (A-25) is approximately given by

$$B_1 \exp(j\delta_1) \doteq -\frac{\rho s}{v} J_1 - B_s \exp(j\delta_s) \qquad \text{(A-26)}$$

$$\doteq -B_s \exp(j\delta_s) . \qquad \text{(A-27)}$$

The content of Eq. (A-27) is the same as that of Eq. (A-22). At the exit end, similarly to Eq. (A-21) we get

$$b\bigg|_{x=L} = \frac{\rho s}{v} j_2 \bigg|_{x=L} \qquad \text{(A-28)}$$

and

$$b\bigg|_{x=L} = \Bigg[B_s \exp(j\theta_s)$$

$$+ B_1 \exp\left(-\frac{L}{\alpha_1}\right) \exp\left\{ j\left(-\frac{\pi L}{\tau_e} + \delta_1\right)\right\}$$

$$+ B_2 \exp(j\delta_2) \Bigg] \exp(j\omega t) . \qquad \text{(A-29)}$$

From Eqs. (A-23), (A-27) and (A-28) we get

$$B_2 \exp(j\delta_2) = -\frac{\left\{ 1 + \dfrac{\rho_0 g}{v \mu_0}\left(\dfrac{1}{\alpha_1} + j\dfrac{\pi}{\tau_e}\right)\right\}}{1 - \dfrac{\rho g}{v \mu_0}\left(\dfrac{1}{\alpha_2} + j\dfrac{\pi}{\tau_e}\right)}$$

$$\times \left\{ 1 - \exp\left(-\frac{L}{\alpha_1}\right) \exp\left(-j\frac{\pi L}{\tau_e}\right)\right\} B_1 \exp(j\delta_1) . \qquad \text{(A-30)}$$

Magnetic Flux Distribution Curves without Fringing

In the model of the linear induction motor for the two-dimensional analysis in Fig. 34 in Chapter 9, the primary iron core is extended to $\pm \infty$ in the x-direction. The end effect is caused then only by the ends of the primary winding and may be different from the actual end effect, which is caused by the ends of both the primary winding and primary iron core. The difference, however, is small. One example of the magnetic-flux-density-distribution curves calculated by two-dimensional analysis is shown in Fig. 39. As can be seen from this example, the magnetic field excited in zone I ($x < 0$) is very weak and its existence is limited to the small vicinity near the entry end. The fringing of the field is of the order of that near the entry end of the primary iron core. On the contrary, in zone III ($L < x$) the magnetic field exists over a longer distance from the exit end and is very different from the actual magnetic field. However, in zone III there is no primary current in the primary iron core and hence thrust is not generated by the air-gap magnetic field, which would generate a little heat loss in the secondary sheet. The overall influence of the magnetic field in zone III is small and can be neglected in most practical cases.

In actual linear induction motors, which have ends of both the primary winding and the primary iron core, fringing of the magnetic field occurs and a weak magnetic field exists in zone I and zone III. However, the fringing magnetic field can be neglected in most practical cases and is neglected in our one-dimensional analysis. The condition of no fringing can be established by assuming that super-conductive conductors are provided at both entry and exit ends, as shown in Fig. A-7. As is well known, super-conductive conductors have practically no electric resistance and are diamagnetic (Meissner effect). No magnetic flux can cross the super-conductive conductors at the entry and exit ends.

An analysis will be made of the case where the super-conductive conductors are provided at the exit boundary. Let the current in the super-conductive conductors at the exit be denoted by J_s, then its Fourier transform, which corresponds to $\tilde{J}$ of Eq. (104), is given by

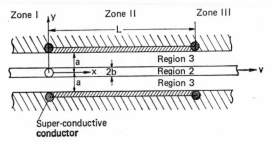

Fig. A-7. Model for the two-dimensional analysis of a linear induction motor without fringing at entry and exit ends.

$$\tilde{J} = J_{\mathrm{s}} \exp{(j\omega L)}. \tag{A-31}$$

Vector potentials produced in the air gap by current J_{s} are given by Eq. (115) in which J_1 is replaced with $\tilde{J}_{\mathrm{s}}$ of Eq. (A-31). Thus we have

$$\phi_{\mathrm{s3}}(x, y) = \frac{1}{2\pi} \int_{-\infty}^{\infty} \mu_3 J_{\mathrm{s}} \frac{G(\xi, y)}{H(\xi)} \exp\{-j\xi(L-x)d\xi . \tag{A-32}$$

In applying the residue theorem to Eq. (A-32), the path of integration in Fig. 35 must be the lower semi-circle for $L \gg x$ and the upper semi-circle for $L < x$. Thus we obtain the following vector potentials:
In zone I ($x < 0$)

$$\phi_{\mathrm{s3}}(x, y) = \mu_3 J_{\mathrm{s}} \frac{G(\xi_0', y)}{H'(\xi_0')} \exp\{-j\xi_0'(L-x)\}. \tag{A-33}$$

In zone II ($0 < x < 1$)

$$\phi_{\mathrm{s3}}(x, y) = \mu_3 J_{\mathrm{s}} \frac{G(\xi_0', y)}{H'(\xi_0')} \exp\{-j\xi_0'(L-x)\}. \tag{A-34}$$

In zone III ($L < x$)

$$\phi_{\mathrm{s3}}(x, y) = \mu_3 J_3 \frac{G(\xi_0, y)}{H'(\xi_0)} \exp\{-j\xi_0(L-x)\}. \tag{A-35}$$

The three vector potentials of Eqs. (A-33), (A-34) and (A-35) are added to the corresponding vector potentials of Eqs. (126), (127) and (128). In order that no field exists in zone III, the sum of Eq. (A-35) and Eq. (128) must be zero everywhere in zone III. Thus we have

$$J_{\mathrm{s}} = J_1 \frac{\exp(jkL) - \exp(j\xi_0 L)}{\xi_0 + k} . \tag{A-36}$$

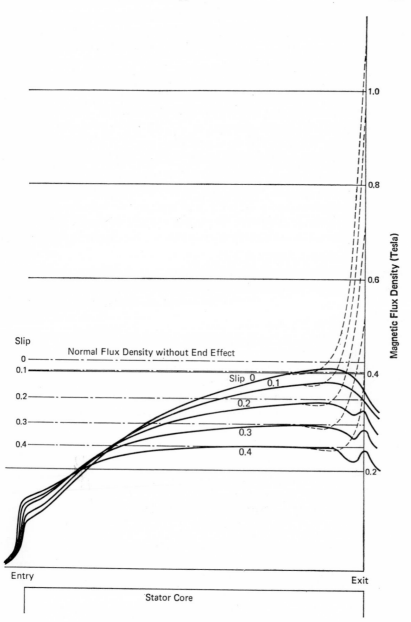

Fig. A-8. Calculated magnetic flux density curves.

This is the current flowing in the super-conductive conductors which eliminates any field entering zone III. The vector potential produced by current J_S in zone II is given by Eqs. (A-34) and (A-36) as follows:

$$\phi_{S3}(x, y) = \mu_3 J_1 \frac{\exp(jkL) - \exp(j\xi_0 L)}{\xi_0 + k} \frac{G(\xi_0', y)}{H'(\xi_0')} \exp\{-j\xi_0'(L-x)\}. \quad \text{(A-37)}$$

The vector potential in zone II $(0 < x < L)$ under the condition of no fringing at the exit end is obtained by adding $\phi_{S3}(x, y)$ of Eq. (A-37) to $\phi_3(x, y)$ of Eq. (127). ϕ_{S3} of Eq. (A-37) has the factor $\exp(j\xi_0'x)$ and is therefore the backward travelling wave. Thus the backward travelling wave given by the third term of Eq. (127) is corrected by the addition of $\phi_{S3}(x, y)$ to Eq. (A-27). Figure A-8 shows the calculated magnetic-flux-density-distribution curves for both cases with and without the addition of the correcting terms $\phi_{S3}(x, y)$ in Eq. (A-37). The continuous lines are the same as those in Fig. 31 and the broken lines are the magnetic-flux-density-distribution curves corrected with $\phi_{S3}(x, y)$ of Eq. (A-37) for the case of no fringing at the exit end. There are sharp rises in the vicinity of the exit end if there is no fringing. Fringing occurs in the actual magnetic field in the air gap and its field distribution would be intermediate between the two cases in Fig. A-8, as shown in Fig. A-3, whose curves were numerically calculated for the case where both primary winding and iron core have the same finite length.

The same kind of correction for no fringing is possible for the entry end by providing it with super-conductive conductors. However, the magnetic field in zone I given by Eq. (126) is of the order of the fringing field, and elimination of the field in zone I is not necessary.

A Compensated Linear Induction Motor with a Single-Phase Compensating Winding or a Concentrated Compensating Winding.

The compensated linear induction motor shown in Fig. 57 has a three-phase compensating winding. The compensating winding may be a single-phase winding. In this case the current in the compensating winding is expressed by

$$j_C = J_C \exp(j\omega t) \quad (A/m). \tag{A-38}$$

The current j_C exists over length $x = -L_C$ and $x = 0$, and then its Fourier transform is given by

$$\tilde{J}_C = \int_{-L_C}^{0} J_C \exp(-j\xi x)dx = \frac{jJ_C}{\xi}\{1 - \exp(j\xi L_C)\}, \tag{A-39}$$

where $\exp(j\omega t)$ is omitted. Replacing $\tilde{J}_1$ in Eq. (115), we have

$$\phi_3(x, y) = \frac{j\mu_3 J_C}{2\pi} \int_{-\infty}^{\infty} \frac{1 - \exp(j\xi L_C)}{\xi} \frac{G(\xi, y)}{H(\xi)} \exp(j\xi x)d\xi. \tag{A-40}$$

Applying the residue theorem to Eq. (A-40), the path of integration in Fig. 35 should be the upper semi-circle for zone II $(0 < x < L)$, and the solution is as follows:

$$\phi_3(x, y) = \mu_3 J_C \frac{1 - \exp(j\xi_0 L_C)}{\xi_0} \frac{G(\xi_0, y)}{H'(\xi_0)} \exp(j\xi_0 x). \tag{A-41}$$

The compensation condition is satisfied if the sum of $\phi_3(x, y)$ in Eq. (A-41) and the second term of Eq. (189) is zero, irrespective of x, and we get

$$\frac{J_A}{\xi_0 + k_A} = \frac{J_C\{\exp(j\xi_0 L_C) - 1\}}{\xi_0}. \tag{A-42}$$

Comparing the current density J_C in Eq. (192) for the three-phase compensating winding, J_C in Eq. (A-42) is $|\xi_0/(\xi_0 + k_C)|$ times higher. Taking ξ_0 in Eq. (194), $|\xi_0/(\xi_0 + k_C)| \doteq \tau_C/(\tau_C - \tau_A)$ for the smaller slip. This ratio is three for $\tau_C/\tau_A = 1.5$, and this means that the single-phase compensating winding needs a current density τ_C at least three times higher than that of a three-phase compensating winding of the same length L_C, and therefore the single-phase compensating winding should be ruled out for practical application.

The compensating winding may be a concentrated winding located at the entry end with a single-phase current flowing. Then the length L_C of the compensating winding in Fig. 57 becomes infinitesimally small under the condition

$$\lim_{L_C \to 0} J_C L_C = J_{CC} \quad (A). \qquad (A\text{-}43)$$

The Fouriers transform for this case is given by Eq. (A-39), making L_C infinitesimally small. Thus we have

$$\tilde{J}_{CC} = \lim_{L_C \to 0} \frac{jJ_C}{\xi} \{1 - \exp(j\xi L_C)\}$$

$$= \lim_{L_C \to 0} \frac{jJ_C}{\xi} \{1 - (1 + j\xi L_C)\}$$

$$= \lim_{L_C \to 0} J_C L_C = J_{CC} .$$

$$(A\text{-}44)$$

Replacing $\tilde{J}_1$ in Eq. (115) with $\tilde{J}_{CC}$ in Eq. (A-44), we have

$$\phi_3(x, y) = \frac{j\mu_3 J_{CC}}{2\pi} \int_{-\infty}^{\infty} \frac{G(\xi, y)}{H(\xi)} \exp(j\xi x)d\xi . \qquad (A\text{-}45)$$

Applying the residue theorem to Eq. (A-45), the path of integration in Fig. 35 should be the upper semi-circle for zone II $(0 < x < L)$, and the solution is as follows:

$$\phi_3(x, y) = \mu_3 J_{CC} \frac{G(\xi_0, y)}{H'(\xi_0)} \exp(j\xi_0 x) . \qquad (A\text{-}46)$$

The compensation condition is satisfied if the sum of $\phi_0(x, y)$ in Eq. (A-46) and the second term of Eq. (189), which gives the entry-end-effect wave, is zero, irrespective of x, and we get

$$J_{CC} = -\frac{J_A}{\xi_0 + k_A} . \qquad (A\text{-}47)$$

Comparing the current density J_C of the three-phase compensating winding in Eq. (196), the current J_{CC} of the concentrated compensating winding is given by

$$J_{\text{cc}} = -\frac{2J_{\text{c}}}{\xi_0 + k_{\text{c}}}. \tag{A-41}$$

ξ_0 is approximated by Eq. (194), which is expressed in terms of the three-phase compensated winding as follows:

$$\xi_0 = -\frac{\pi(1 - s_{\text{c}})}{\tau_{\text{c}}}, \tag{A-42}$$

where s_{c} is slip with respect to the three-phase compensating winding. From Eqs. (A-41) and (A-42) we get

$$J_{\text{cc}} = \frac{2\tau_{\text{c}} J_{\text{c}}(1 - s_{\text{c}})}{\pi s_{\text{c}}}. \tag{A-43}$$

For $s_{\text{c}} = 0.25$ we have $J_{\text{cc}} = 1.9\tau_{\text{c}} J_{\text{c}}$, which is by 5% less than the total ampere turn $2\tau_{\text{c}} J_{\text{c}}$ of the three-phase compensating winding. However it should be pointed out that the concentrated compensating winding does not produce positive thrust while the three-phase compensating winding does. It may be concluded then that the three-phase compensating winding in Chapter 13 is much more economical.

Appendix VI

A Table of Comparison of Linear Induction Motors

	Motor A[a]	Motor B	Motor C	Motor D	Motor E[a]
Frequency (Hz)	50	50	150	50	0–175
Synchronous speed (m/s)	9	150	150	150	70
Number of poles	4	4	10	8	6
Gap length (mm)	15	40	40	40	28
Secondary sheet[b]	5-mm thick	10-mm thick	10-mm thick	10-mm thick	6-mm thick
Core dimension (mm × mm)	360 × 90	6000 × 1000	5000 × 500	15000 × 1000	1200 × 150
Primary resistance (Ω)	0.53	1.6×10^{-2}	4.6×10^{-2}		0.083
Primary leakage reactance (Ω)	1.91	8×10	6.9×10^{-1}		2.49

[a] Motor A and motor E were actually built and tested.
[b] Aluminum.

Nomenclature

a	Half-gap length
$\mathbf{B}$	Vector for magnetic flux density
B_1	Complex amplitude of magnetic flux density for entry-end-effect wave
B_2	Complex amplitude of magnetic flux density for exit-end-effect wave
$B(t)$	t-function factor for b
$B(x)$	x-function factor for b
b	Magnetic flux density, half thickness of secondary sheet
b_e	End-effect component of magnetic flux density in air gap
b_S	Steady state component of magnetic flux density in air gap
c	Half width of secondary sheet
D	Stack length of primary core
$\mathbf{D}$	Vector for dielectric flux density
d	Half stack length of primary iron core
E_1	Primary electromotive force per phase
$\mathbf{E}$	Vector for electric field intensity
e_u, e_v, e_w	Induced electromotive forces in primary winding of balanced linear induction motor
e_u', e_v', e_w'	Unbalanced induced electromotive forces in primary winding under influence of end effect
$e_{us}', e_{vs}', e_{ws}'$	Induced electromotive forces in primary winding without end effect
e_2	Induced electromotive force per meter in secondary conductor
F	Thrust, $F = F_n + F_e$
F_n	Normal thrust, see Eq. (58)
F_e	Thrust, see Eqs. (59) and (60)
F_{inst}	Instantaneous thrust
f	Frequency
g	Gap length, including nonmagnetic secondary sheet
H	Magnetic field intensity
$\mathbf{H}$	Vector for magnetic field intensity
I_1, i_1	Actual primary current
i_2	Actual secondary current

i_u, i_v, i_w	Three-phase currents of primary winding
$\mathbf{J}$	Vector for current
J_1, j_1	Equivalent primary current sheet
J_2, j_2	Current of secondary conductive sheet
k	Travelling constant of travelling wave
k_1	Travelling constant of entry-end-effect wave
k_2	Travelling constant of exit-end-effect wave
k_{w1}	Winding coefficient of primary winding
L	Londitudinal length of primary iron core
L_A	Length of primary winding A
L_B	Length of primary winding B
L_C	Length of compensating winding
m_1	Number of phases of primary winding
m_2	Number of phases of secondary winding
P	Number of poles
P_A, P_B	Number of poles of winding A and winding B, respectively
p_1	Number of pole pairs for primary
p_2	Number of pole pairs for secondary
q	Number of coil sides per phase per pole
R	Resistance of secondary winding
s	Slip
t	Time
$\mathbf{V}$	Vector for velocity
v	Motor speed
v_e	Speed of end-effect wave
v_S	Synchronous speed
w_1	Number of turns per phase for primary winding
w_2	Number of turns per phase for secondary winding
w_C	Number of turns per coil
X	Real part of Eq. (29)
Y	Imaginary part of Eq. (29)
Z_1	Primary leakage impedance
Z_2	Secondary leakage reactance
Z_m	Magnetizing impedance
Z_{m2}	Impedance of balanced linear induction motor when primary leakage impedance is zero

Greek letters

α_1	Length of penetration of entry-end-effect wave
α_2	Length of penetration of exit-end-effect wave
β	Short pitch factor, real part of ξ_0 and ξ_0'
γ	Given by Eq. (96)
δ_1, δ_1'	Phase angle of B_1 wave

$\bar{\delta}_S$	Phase angle of B_S wave
ε	Base of natural logarithm
λ	Flux linkage of primary winding per pole
λ_C	Flux linkage per coil of primary winding
$\lambda_a, \lambda_b, \lambda_c$	Flux linkage of phase a, b and c, respectively
μ	Permeability
μ_0	Permeability of air
μ_1	Permeability of primary iron core
μ_2	Permeability of secondary conductor
μ_3	Permeability of air gap $(=\mu_0)$
ξ	Variable for Fourier transform
ξ_0	Characteristic root given by Eqs. (121) and (122) (travelling constant for entry-end-effect wave)
ξ_0'	Characteristic root given by Eqs. (121) and (122) (travelling constant for exit-end-effect wave)
ρ	Resistivity
ρ_S	Surface resistivity of secondary conductive sheet
σ	Conductivity
σ_2	Conductivity of secondary conductor
τ	Pole pitch, half-wave length
τ_A, τ_B	Pole pitch of winding A and winding B
τ_C	Pole pitch of compensating winding
τ_e	Half-wave length of end-effect wave
$\boldsymbol{\Phi}$	Vector potential
ϕ	z component of vector potential
ϕ_1	Vector potential in primary iron core
ϕ_2	Vector potential in secondary sheet
ϕ_3	Vector potential in air gap
ω	Angular frequency of power supply

References

1. E. R. Laithwaite, "Induction Machines for Special Purposes," London, George Newness, Ltd., 1966.
2. S. Yamamura and E. Ono, *Linear synchronous motor*, Report 371 of the General Meeting, Tokyo Section, IEE Japan, 1967.
3. S. Yamamura, H. Ito and F. Ahmed, *Winding of linear motors*, Report 491 of the General Meeting, Tokyo Section, IEE Japan, 1968.
4. S. Yamamura, H. Ito and F. Ahmed, *End effect of linear induction motors*, Report 145 of the General Meeting, Tokyo Section, IEE Japan, 1968.
5. S. Yamamura, F. Ahmed and Y. Ishikawa, *Magnetic flux distribution and thrust of a two-sided linear induction motor with a large air gap*, Report 1444 of the General Meeting, Tokyo Section, IEE Japan, 1968.
6. S. Yamamura, H. Ito and F. Ahmed, *End effect of linear induction motors*, Report 570 of the Joint Meeting of Four Electrical Institutes in Japan, 1969.
7. S. Yamamura and H. Ishikawa, *Determination of magnetic flux distribution of linear induction motors by means of the relaxation method*, Report 571 of the Joint General Meeting of Four Electrical Institutes in Japan, 1969.
8. S. Yamamura, Y. Ishikawa and H. Ito, *Performance calculation of linear induction motors by means of the relaxation method*, Report 234 of the Joint General Meeting of Four Electrical Institutes in Japan, 1969.
9. S. Yamamura and H. Ito, *End effect of linear induction motors*, Report 511 of the Joint General Meeting of Four Electrical Institutes in Japan, 1969.
10. S. Yamamura, Y. Ishikawa and H. Ito, *Performance calculation of linear induction motors by means of the relaxation method*, Report 570 of the Joint General Meeting of Four Electrical Institutes in Japan, 1969.
11. S. Yamamura and H. Ito, *Performance of high-speed linear induction motors*, Report 138 of the Joint General Meeting, Tokyo Section, IEE Japan, 1970.
12. S. Yamamura and Y. Ishikawa, *End effect of linear induction motors with an iron sheet secondary*, Report 137 of the General Meeting, Tokyo Section, IEE Japan, 1970.
13. S. Yamamura, H. Ito and F. Ahmed, *End effect of linear induction motors*, Jour. IEE Japan, March, 1968, p. 459.
14. S. Yamamura, H. Ito and Y. Ishikawa, *Influence of the end effect on the performance of linear induction motors*, Jour. IEE Japan, Feb., 1971, p. 309.

15. S. Yamamura, H. Ito and Y. Ishikawa, *Compensation of the end effect of the linear induction motor*, Symposium Report IEE Japan, 1971.
16. S. Yamamura, H. Ito and Y. Ishikawa, *Influence of the end effect of linear induction motors and its compensation*, Report RM-71-1 of the Study Committee on Rotating Machinery, IEE Japan, Feb., 1971.
17. H. Bolton, *Transverse edge effect in short-rotor induction motors*, Proc. IEE, Vol. 116, May, 1968, p. 725.
18. S. Nonaka and K. Yoshida, *Analysis of the characteristics of the two-sided linear induction motor*, Jour. IEE Japan, Vol. 90, No. 5, 1970, p. 880.
19. S. Yamamura and H. Ito, *End effect of a linear induction motor of the wound-secondary type*, Report 434 of the General Meeting, IEE Japan, 1971.
20. Office of High Speed Ground Transportation, U.S.A., *Study of linear induction motor and its feasibility for high speed ground transportation*, June, 1967.
21. S. Yamamura and Y. Ishikawa, *Lateral force of linear induction motors under the influence of the end effect*, Report 435 of the General Meeting, IEE Japan, 1971.
22. S. Yamamura, *End effect of linear induction motors and its compensation*, Symposium report S 4-2, of the General Meeting, IEE Japan, 1971.
23. S. Yamamura, H. Ito and Y. Ishikawa, *Influence of the end effect on linear induction motor performance and its compensation*, Jour. IEE Japan, Nov., 1971, p. 2075.
24. S. Yamamura, Y. Ishikawa and H. Ito, *Calculation and measurement of high-speed induction motor performance*, Report of the Tokyo Section, IEE Japan, Nov., 1971.
25. S. Yamamura, H. Ito and Y. Ishikawa, *Influence of the end effect on linear induction motor performance and its compensation*, Report of the Joint Meeting of Four Electrical Institutes in Japan, Oct., 1971.
26. Y. Ishikawa, *Linear Induction Motor*, Master's thesis, Electrical Engineering Department, Faculty of Engineering, University of Tokyo, March, 1970.
27. S. Yamamura, H. Ito and Y. Ishikawa, *Theories of the linear induction motor and compensated linear induction motor*, Transactions Paper T72, 060-7, IEEE, Feb. 1972.
28. S. Yamamura, H. Ito and Y. Ishikawa, *Theory of the linear induction motor and compensated linear induction motor*, IEEE Transaction on Power Apparatus and Systems, July/August 1972, p. 1700.

DATE DUE

FEB			
DEC 4 '89			
GAYLORD			PRINTED IN U.S.A